Artificial Intelligence for Transportation

Transforming Mobility and Infrastructure

Editors:

Dhanashree Kulkarni
Department of Computer Science and Engineering,
Angadi Institute of Technology and Management,
Affiliated to Visvesvaraya Technological University (VTU),
Belagavi, Karnataka, India

Anand B. Deshpande
Principal & Director,
Angadi Institute of Technology and Management,
Affiliated to Visvesvaraya Technological University (VTU),
Belagavi, Karnataka, India.

Vania V. Estrela
Department of Telecommunications
Federal Fluminense University, Rio de Janeiro, Brazil

CRC Press
Taylor & Francis Group
Boca Raton London New York

CRC Press is an imprint of the
Taylor & Francis Group, an **informa** business
A SCIENCE PUBLISHERS BOOK

First edition published 2027
by CRC Press
2385 NW Executive Center Drive, Suite 320, Boca Raton FL 33431

and by CRC Press
4 Park Square, Milton Park, Abingdon, Oxon, OX14 4RN

CRC Press is an imprint of Taylor & Francis Group, LLC

Library of Congress Cataloging-in-Publication Data (applied for)

ISBN: 978-1-041-07186-0 (hbk)
ISBN: 978-1-041-09037-3 (pbk)
ISBN: 978-1-003-64814-7 (ebk)

DOI: 10.1201/9781003648147

Typeset in Times New Roman
by Prime Publishing Services

Preface

The present book presents a comprehensive exploration of how artificial intelligence (AI) is reshaping the way we move people and goods, while redefining transportation infrastructure for the future. It emphasizes the transformative role of AI technologies—such as machine learning, computer vision, natural language processing, and data analytics—in building smarter, safer, and more sustainable transportation systems.

This book presents a comprehensive exploration of how emerging AI technologies are reshaping the future of mobility. The book spans foundational concepts to cutting-edge innovations, beginning with the transformative role of AI in modern transportation systems and the evolution of Intelligent Transportation Systems. The text delves into specialized domains such as AI-powered driving analytics, intelligent traffic management, and augmented reality–enabled safety navigation. It further highlights next-generation breakthroughs, including 6G-integrated mobility, machine intelligence for autonomous vehicles, edge-AI frameworks, and quantum computing. Pushing into the frontier of future mobility, generative AI–based approaches for road infrastructure design and robust AI foundations for vehicle-to-vehicle communication underscore the book's focus on building resilient, efficient, and intelligent transportation ecosystems. Serving as a vital reference for researchers, industry professionals, and policymakers, this book offers a holistic view of how AI is shaping safer, smarter, and more sustainable transportation for the future.

Ultimately, this book provides a forward-looking vision of AI-enabled transportation that enhances mobility, reduces emissions, optimizes infrastructure planning, and contributes to the broader development of smart cities. An essential reference for students, researchers, engineers, urban planners, policymakers, and industry professionals and leaders seeking to understand and leverage AI's potential to transform global transportation ecosystems, this volume is both a practical guide and a visionary outlook.

We extend our gratitude to the contributors for their invaluable insights and expertise. We hope this book will inspire and inform, fostering a deeper understanding of the transformative potential embedded in AI in transportation. We also hope that this book will motivate and support those working in the rapidly advancing domain of AI for transportation, and encourage worldwide discussions on nourishing a growing global population while preserving our planet.

Dr Dhanashree Kulkarni
Dr Anand B. Deshpande
Dr Vania V. Estrela

Acknowledgement

The development of this book has been a truly collaborative endeavor that transcended geographical boundaries and united the collective expertise of distinguished institutions. It is with profound gratitude that we acknowledge the contributions of the organizations and individuals whose support was instrumental in bringing this work to fruition.

We express our deep appreciation for Dr Spoorti R. Angadi (Patil), Director, Suresh Angadi Education Foundation (SAEF); Mr Raju Joshi, Administrator, SAEF; and the faculty and staff of SAEF's Angadi Institute of Technology and Management, Belagavi, Karnataka, India, affiliated with Visvesvaraya Technological University (VTU), Belagavi. Their unwavering support and steadfast dedication to fostering innovation in technology and management have significantly enriched the depth and scope of this book. We also extend our sincere thanks to Prof. Priyanka Pujari and Prof. Vishalkirthi Patil for their invaluable assistance and noteworthy contributions to the design and conceptual framing of this work.

We further extend our heartfelt appreciation to Fluminense Federal University (UFF), Brazil, for its enduring commitment to academic excellence and its significant advancements in the field of agriculture. The vibrant scholarly environment nurtured by UFF has greatly contributed to the intellectual richness reflected in these pages. The collaborative spirit embodied by these institutions underscores a shared pursuit of advancing knowledge and addressing the evolving challenges of the twenty-first century. This book stands as a testament to the transformative potential of international cooperation and the remarkable outcomes that emerge when scholars and institutions across the globe unite in pursuit of a common vision.

We express our gratitude to all contributors whose expertise, dedication, and thoughtful insights have shaped this work. We also express our sincere gratitude to our respective families whose constant support, encouragement, and understanding made it possible for us to devote our time and energy to this endeavor. Their patience and goodwill have been invaluable pillars throughout the development of this book. Together, we have laid a foundation for discourse that transcends borders and fosters sustainable, innovative pathways for the future of transportation systems.

Dr Dhanashree Kulkarni
Dr. Anand B. Deshpande
Dr. Vania V. Estrela

Contents

List of Contributors

Aditya Patil
Department of Computer Science and Engineering, Jain College of Engineering and Research, Belagavi, Karnataka, India.

Akshay Pujeri
Department of Computer Science and Engineering, Angadi Institute of Technology and Management, Affiliated to Visvesvaraya Technological University (VTU), Belagavi, Karnataka, India.

Alok Gaddi
KLE Technological University, School of Management Studies & Research (SMSR), Hubli, Karnataka, India.

Anand Deshpande
Department of Artificial Intelligence and Data Science, Angadi Institute of Technology and Management, Belagavi, Karnataka, (Affiliated to Visvesvaraya Technological University (VTU)), India.

Anand Jumnal
Department of Computer Science & Engineering, BLDEA's V. P. Dr. P. G. Halakatti College of Engineering and Technology, Vijayapura, Karnataka.

Angela Maria Alves
Dom Pedro I Highway (SP-65), Km 143,6 - Chácaras Campos dos Amarais, Campinas - SP, 13069-901, Brazil.

Anjanabhargavi Kulkarni
Department of Computer Science and Engineering, Visvesvaraya Technological University, Belagavi, Karnataka, India.

Arun Tigadi
Department of Electronics and Communication Engineering, KLE Technological University Dr. M. S. Sheshgiri Campus, Belagavi, Karnataka, India.

Asiya Khan
University of Plymouth, Plymouth, United Kingdom.

Bhanuteja G.
School of Electronics and Communication Engineering, REVA University, Bengaluru, Karnataka, India.

Dhanashree Kulkarni
Department of Computer Science & Engineering, Angadi Institute of Technology and Management, Belagavi, (Affiliated to Visvesvaraya Technological University (VTU)), Karnataka, India.

Euclides Lourenco Chuma
Paschoal Marmo, 1888 - Jardim Nova Itália, Limeira - SP, 13484-332, Brazil.

Gabriel C. Vaz
School of Electrical and Computer Engineering, State University of Campinas, Campinas, Brazil.

Gabriel G. de Oliveira
School of Electrical and Computer Engineering, State University of Campinas, Campinas, Brazil.

Gautam Arjun Dematti
Department of Computer Science and Engineering, Angadi Institute of Technology and Management, Belagavi, Karnataka, India. (Affiliated to Visvesvaraya Technological University, Belagavi, Karnataka, India.)

Gujanatti Rudrappa
Department of Electronics and Communication Engineering, KLE Technological University Dr. M. S. Sheshgiri Campus, Belagavi, Karnataka, India.

Gourav Vivek Kulkarni
Department of Mechanical Engineering, KLS Gogte Institute of Technology, Belagavi, Karnataka, India.

Julie Stephany Berrio Perez
Australian Centre for robotics, University of Sydney, Australia.

Kartik Cholachgudda
School of Electronics and Communication Engineering, REVA University, Bengaluru, Karnataka, India.

Karuna C. Gull
CSBS Department, SGBIT, Visvesvaraya Technological University, Belagavi, Karnataka, India.

Kuldeep Sambrekar
Department of Computer Science and Engineering (AI & ML), KLS Gogte Institute of Technology, Belagavi, Karnataka, India.

Manjunath Patil
Department of Computer Science & Engineering, Angadi Institute of Technology and Management, Belagavi, Karnataka.

Pradeep Angadi
National Oilwell Varco, Greengate, Manchester, United Kingdom.

Prakash Sonwalkar
Department of Computer Science and Engineering (AIML), Jain College of Engineering and Research, Affiliated to Visvesvaraya Technological University (VTU), Belagavi, Karnataka, India.

Pranav Kamble
Department of Computer Science and Engineering, Angadi Institute of Technology and Management, Affiliated to Visvesvaraya Technological University (VTU), Belagavi, Karnataka, India.

Prasad Kulkarni
Jain College Engineering & Research, Belagavi, Karnataka, India

Prathmesh Redekar
Department of Computer Science and Engineering (AIML), Jain College of Engineering and Research, Affiliated to Visvesvaraya Technological University (VTU), Belagavi, Karnataka, India.

Prathvi Teli
Department of Computer Science and Engineering, Angadi Institute of Technology and Management, Affiliated to Visvesvaraya Technological University (VTU), Belagavi, Karnataka, India.

Priyanka Pujari
Department of Computer Science and Engineering, Angadi Institute of Technology and Management, (Affiliated to Visvesvaraya Technological University (VTU)), Belagavi, Karnataka, India

R. H. Goudar
Department of Computer Science and Engineering, Visvesvaraya Technological University, Belagavi, Karnataka, India.

Ravi Shankar D.
School of Electronics and Communication Engineering, REVA University, Bengaluru, Karnataka, India.

Rohit Patel
Department of Computer Science and Engineering, Angadi Institute of Technology and Management, Affiliated to Visvesvaraya Technological University (VTU), Belagavi, Karnataka, India.

Ruturaj Patil
Department of Management Studies, Visvesvaraya Technological University, Belagavi, Karnataka, India.

Sachin Urabinahatti
Department of Artificial Intelligence and Machine Learning, BMS Institute of Technology and Management, Bengaluru, Karnataka, India.

Sakshi Hatti
Department of Computer Science and Engineering, Jain College of Engineering and Research, Belagavi, Karnataka, India.

Samiksha Shivaji Mayekar
Department of Mechanical Engineering, KLS Gogte Institute of Technology, Belagavi, Karnataka, India.

Shalini L.
Department of Computer Science, Sahyadri College of Engineering and Management, Mangaluru, India.

Shivanand S. Kagawade
Department of MBA, SAEF's Angadi Institute of Technology & Management, Belagavi, Karnataka, India.

Shivashankar K.
Karnataka Folklore University, Shiggaon, Gotagudi, Karnataka, India.

Sonam Bhandurge
Department of Computer Science and Engineering, Gogte Institute of Technology, Belagavi, Karnataka, India.

Srividya Bhat
Department of Computer Science and Engineering, Sahyadri College of Engineering and Management, Mangaluru, Karnataka, India.

Stephany Berrio Perez
Department of Australian centre of field robotics, University of Sydney, Sydney, Australia.

Supriya B.
Department of Computer Science and Engineering, Sahyadri College of Engineering and Management, Mangaluru, Karnataka, India.

Sushant Jadhav
Department of Electronics and Communication Engineering, KLE Technological University Dr. M. S. Sheshgiri Campus, Belagavi, Karnataka, India.

Tanishq Jain
Department of Computer Science and Engineering, KIT's College of Engineering, Kolhapur, Maharashtra, India.

Uttam Patil
Department of CSE, Jain College of Engineering, Belagavi, Karnataka, India.

V. Estrela
Telecommunications Department, Fluminense Federal University, Rio de Janeiro, Brazil.

Vania V. Estrela
Telecommunications Department, Fluminense Federal University, Rio de Janeiro, Brazil

Vitor Eduardo Molina Junior
Faculty of Technology - Limeira, Brazil.

William Machado Emiliano
Faculty of Technology - Limeira, Brazil.

Yash Mulay
Department of Computer Science and Engineering, KIT's College of Engineering, Kolhapur, Maharashtra, India.

Yash Tanaji Shinde
Department of Computer Science and Engineering, Jain College of Engineering and Research, Affiliated to Visvesvaraya Technological University, Belagavi, Karnataka, India.

Yuri Alexandre Meyer
Faculty of Technology - Limeira, Brazil.

Yuzo Iano
School of Electrical and Computer Engineering, State University of Campinas, Campinas, Brazil.

1 Role of Artificial Intelligence in Transportation Systems

Dhanashree Kulkarni[1*], Anand Deshpande[2], V. Estrela[3] and Julie Stephany Berrio Perez[4]

Artificial intelligence (AI) has emerged as a transformative force within intelligent transportation systems (ITS), reshaping how people and goods move across modern mobility networks. This chapter explores the pivotal role of AI in enhancing transportation efficiency, safety, and sustainability. It begins by examining the need to adopt AI in transportation, highlighting its ability to improve operational efficiency, reduce environmental impact, support smarter traffic management, and strengthen safety for drivers and pedestrians. The chapter then discusses key applications of AI, including autonomous vehicles, intelligent routing, demand forecasting, fleet management, and real-time incident detection. Additionally, it outlines the various modes of transportation—road, rail, air, maritime, and public transit—where AI-enabled ITS technologies are implemented to optimize operations and enhance user experience. By integrating advanced data analytics, machine learning, and automated decision-making, AI significantly elevates the capabilities of traditional transportation systems. This chapter provides a comprehensive understanding of how AI-driven ITS solutions are shaping the future of global mobility and enabling more intelligent, resilient, and sustainable transport infrastructures.

1. Introduction

Artificial intelligence (AI) has emerged as a foundational technological paradigm that is redefining the capabilities of modern computing systems. At its core, AI facilitates sophisticated functions such as perception, adaptive learning, strategic decision-making, and contextual understanding, thereby supporting complex problem-solving in dynamic environments. AI techniques have been increasingly integrated into a wide spectrum of application domains, including cybersecurity, healthcare diagnostics, financial forecasting and fraud detection, autonomous robotics, smart manufacturing, natural language processing, textual analytics, and

1 Department of Computer Science & Engineering, Angadi Institute of Technology and Management, Belagavi, Affiliated to Visvesvaraya Technological University (VTU), Karnataka, India.
2 Department of Artificial Intelligence and Data Science, Affiliated to Visvesvaraya Technological University (VTU), Angadi Institute of Technology and Management, Belagavi, India.
3 Telecommunications Department, Fluminense Federal University, Rio de Janeiro, Brazil.
4 Australian Centre for robotics, University of Sydney, Australia.
* Corresponding author: dskulkarni10@gmail.com

DOI: 10.1201/9781003648147-01

intelligent automation across industrial processes [1, 2, 3, 4, 5, 6]. Among these diverse applications, a notable contemporary shift is the rising prominence of AI-enabled intelligent transportation systems (ITS).

Artificial intelligence is rapidly transforming the global transportation landscape, emerging as one of the most powerful technologies driving modern mobility [7]. As cities grow, populations increase, and travel demand rises, traditional transportation systems face significant challenges in managing congestion, ensuring safety, and delivering efficient services. AI offers a new level of intelligence and automation that allows transportation networks to operate more effectively, addressing the limitations of conventional systems and creating smarter, more responsive mobility solutions [8].

The growing reliance on digital data—such as GPS devices, mobile apps, traffic cameras, and connected vehicles—provides a foundation for AI-driven innovation [9]. Transportation systems generate massive datasets every second, and AI technologies such as machine learning, deep learning, and computer vision analyze this information to recognize patterns, predict problems, and support real-time decision-making [10]. This data-driven intelligence empowers transportation authorities, mobility service providers, and vehicle manufacturers to plan and operate systems with higher precision and reliability [11].

One of the strongest motivations for adopting AI in transportation is the urgent need to enhance *safety*. Human error remains the primary cause of road accidents worldwide, leading to significant loss of life and property [12, 13]. AI-powered safety systems, such as collision detection, driver monitoring, lane-keeping assistance, and automated braking, help reduce the risk of accidents by reacting faster than human drivers [14]. With advanced perception and prediction capabilities, AI improves the safety of drivers, passengers, pedestrians, and cyclists.

Congestion is another major challenge that modern cities struggle to manage. Traditional traffic control systems use fixed schedules that cannot adapt to changing conditions, resulting in delays and inefficiency. AI-based *intelligent traffic management systems* analyze real-time traffic flow, predict congestion before it occurs, and adjust signal timings dynamically. By optimizing traffic movement, rerouting vehicles, and balancing road usage, AI reduces travel time, fuel consumption, and overall road stress. This smarter traffic management creates more efficient and resilient urban mobility networks.

Artificial intelligence also plays a transformative role in *public transportation* by enhancing planning, operations, and passenger experience. Transit agencies use AI to analyze ridership patterns, optimize bus schedules, predict demand, and improve fleet dispatching. AI-enabled passenger information systems provide real-time updates on arrivals, delays, and route changes, making public transit more reliable and user-friendly [15]. Smart ticketing systems using AI-driven platforms further streamline the travel experience and encourage wider use of public transport.

The logistics and freight sector too benefits significantly from AI. With the rise of e-commerce and global supply chains, the demand for fast, efficient, and reliable delivery has intensified. AI improves route planning, fleet management, cargo tracking, and warehouse automation [16]. Predictive maintenance systems detect potential vehicle issues before breakdowns occur, reducing downtime and cost [17]. Through these capabilities, AI ensures goods are transported more efficiently, safely, and economically.

Artificial intelligence is also a key enabler of *autonomous and connected vehicles*, which represent the future of intelligent mobility. Autonomous vehicles rely on AI algorithms to interpret sensor data, detect obstacles, understand traffic signals, and make real-time driving decisions. Connected vehicle technologies allow vehicles to communicate with each other and with road infrastructure, enhancing situational awareness and reducing the likelihood of accidents. Together, these advancements pave the way for safer, more coordinated, and highly automated transportation systems [18].

The influence of AI extends beyond road transport into rail, aviation, and maritime sectors. Railways use AI for predictive maintenance, scheduling, energy optimization, and automated train control. Aviation benefits from AI in air traffic management, runway safety, navigation systems, and passenger flow optimization [19]. In maritime transport, AI supports vessel routing, port operations, and cargo handling. These applications highlight the versatility of AI across multiple transportation modes, each benefiting from increased efficiency and safety [20].

Environmental sustainability is another major driver of AI adoption in transportation. Transportation is a significant contributor to global greenhouse gas emissions, and AI plays a crucial role in reducing its environmental impact. AI optimizes routes to minimize fuel consumption, reduces idle time through better traffic flow, supports electric vehicle charging management, and promotes eco-driving practices [21]. By enabling greener mobility patterns, AI helps cities achieve sustainability goals and improve public health.

Artificial intelligence is reshaping transportation by introducing automation, intelligence, and efficiency across all modes of travel. It enhances safety for road users, eases congestion through smarter traffic control, modernizes public transportation, strengthens logistics operations, and supports next-generation autonomous mobility. As transportation systems continue to evolve, AI will remain a foundational technology driving innovation, sustainability, and long-term mobility transformation. This introduction provides a holistic understanding of AI's role in transportation and sets the stage for deeper exploration of its applications and impact [22]. Figure 1.1 depicts the rationale for adopting AI in transportation systems.

2. AI in Multiple Modes of Transport

Artificial intelligence in transportation systems integrates advanced communication, sensing, and computational technologies across multiple modes of transport to enhance safety, efficiency, and sustainability. The key transportation modes involved have been discussed in this section.

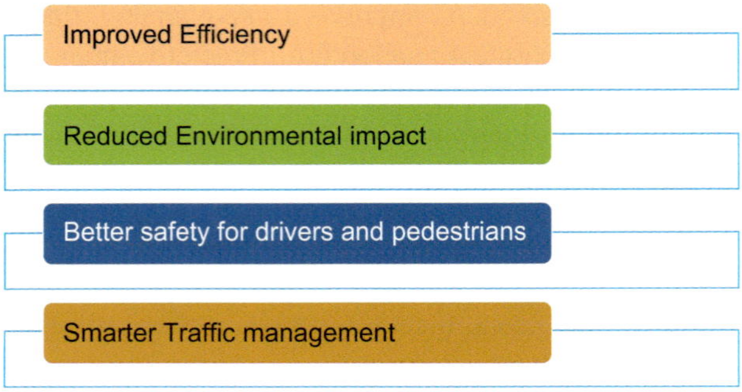

Fig. 1.1 Rationale for adopting AI in transportation systems (created by author)

2.1 Road Transportation

Artificial intelligence is transforming road transportation by improving safety, efficiency, and sustainability. From advanced driver assistance systems (ADAS) to fully autonomous vehicles, AI enables real-time perception, decision-making, and vehicle control. Modern road networks increasingly depend on AI algorithms to manage complex traffic situations, reduce congestion, and adapt dynamically to changing road conditions. As vehicles become more connected and intelligent, the road transport ecosystem is rapidly evolving toward automation and predictive operations. Intelligent traffic management systems use AI to optimize signal timing, identify bottlenecks, and manage roadway capacity more effectively. Machine learning models analyze real-time sensor, GPS, and camera feeds to predict traffic patterns and automatically adjust signal cycles. This leads to reduced congestion, lower travel times, and minimized fuel consumption [23]. AI-powered platforms also support incident detection—spotting accidents or stalled vehicles earlier than human operators and enabling faster emergency response [24]. Autonomous vehicles (AVs) rely heavily on AI technologies such as computer vision, deep learning, and sensor fusion. These systems interpret road signs, detect pedestrians, recognize lane markings, and anticipate the behavior of other road users. Vehicle-to-everything (V2X) communication further enhances safety by enabling cars to communicate with infrastructure, other vehicles, and traffic networks. AI improves route planning, energy usage, and driving comfort, paving the way toward safer and more efficient driverless mobility [25].

2.2 Rail Transportation

Rail systems benefit from AI to ensure reliable operations, predictive maintenance, and optimized scheduling. As rail networks expand and become more complex, AI-driven digitalization enhances safety and asset management. AI models process data from sensors attached to tracks, locomotives, and rolling stock to support continuous monitoring and reduce operational risks [26]. Smart railways use automation to improve punctuality, manage traffic on busy corridors, and enhance passenger experience. AI-powered predictive maintenance helps rail operators identify faults before they lead to delays or accidents. Machine learning algorithms analyze vibration signals, acoustic patterns, temperature changes, and structural conditions to detect potential failures in tracks, wheels, and engines. By predicting wear and tear, AI reduces downtime, avoids disruptions, and extends asset lifespan. Rail operators save significant costs by switching from traditional time-based maintenance to condition-based, data-driven maintenance. Automatic Train Operation (ATO) systems use AI to regulate train speeds, braking, headways, and stopping accuracy. AI enhances existing systems like Automatic Train Control (ATC) and European Train Control System (ETCS) by enabling real-time optimization. With machine learning models, train schedules can be dynamically adjusted based on delays, passenger density, or unexpected disruptions. This ensures smoother operations, higher service frequency, and improved energy efficiency [27].

2.3 Air Transportation

Artificial intelligence plays a significant role in aviation by enhancing safety, optimizing flight operations, and improving airport efficiency. From pilot assistance systems to aircraft health monitoring, AI helps detect anomalies earlier and reduces human error. In air traffic management, AI tools analyze weather patterns, aircraft trajectories, and traffic density to optimize routing and minimize delays [28]. AI-driven decision support systems assist air traffic controllers in managing the increasingly crowded skies. Airports employ AI to streamline security screening, baggage handling, passenger flow management, and gate allocation. Vision-based AI systems improve threat detection at security checkpoints, while machine learning algorithms track luggage movements to minimize misplacement. AI analyzes passenger queues and adjusts staffing levels or routing to maintain efficient terminal operations. For airlines and airports, AI enables better gate scheduling, reducing delays and turnaround times. AI is central to the development of multimodal transportation, where passengers and goods seamlessly shift between road, rail, air, and maritime modes. AI-driven mobility-as-a-service (MaaS) platforms integrate real-time data to provide optimal route combinations for travelers. Autonomous drones, smart logistics hubs, and hyperloop systems also rely on AI for planning, control, and safety. As transportation becomes more interconnected and data-driven, AI will remain the backbone of sustainable, efficient, and resilient mobility ecosystems [29]. A central application in aviation is Air Traffic Management (ATM), which coordinates aircraft movements in airspace and on the ground. ATM uses radar, satellite-based navigation, and communication systems to guide aircraft along efficient routes, manage takeoffs and landings, and prevent conflicts between aircraft. Intelligent ATM improves airspace capacity, reduces delays, and enables more precise flight paths, which in turn lowers fuel consumption and emissions. Advanced decision-support tools help controllers monitor real-time air traffic conditions, detect potential conflicts early, and implement efficient routing strategies to maintain smooth traffic flow. AI also enhances safety and passenger experience through runway safety and collision avoidance systems, as well as smart baggage handling and passenger flow systems. Runway safety systems use sensors, cameras, and warning technologies to detect unauthorized runway incursions and support pilots during landings and takeoffs. Collision avoidance systems such as Traffic Collision Avoidance System (TCAS) help pilots maintain safe separation in the air by giving real-time alerts [28]. On the ground, smart baggage handling systems use RFID, automated sorting, and tracking technologies to minimize lost luggage and speed up processing. Passenger flow systems analyze movement patterns to reduce queues, optimize security checks, and improve boarding efficiency. Together, these intelligent systems create safer, faster, and more seamless aviation operations.

2.4 Water/Maritime Transportation

Artificial intelligence is reshaping the maritime transportation sector by enabling smarter, safer, and more efficient vessel operations. Traditional navigation relies heavily on human experience, but AI-enhanced systems integrate data from GPS, radar, automatic identification system (AIS), sonar, and satellite feeds to support real-time decision-making. These technologies help ships avoid hazards, respond quickly to changing weather conditions, and navigate complex waterways with greater precision. As a result, maritime safety improves, and the risk of collisions, groundings, and operational errors is significantly reduced [30]. AI also plays an important role in predictive maintenance and vessel health monitoring. Machine-learning models analyze engine performance, fuel usage, vibration patterns, and mechanical wear to detect faults before they occur. This minimizes unexpected equipment failures, reduces downtime, and enhances overall fleet reliability. By scheduling timely maintenance and optimizing spare-part management, shipping companies can cut operational costs and extend the lifespan of critical on-board systems [31].

In maritime logistics, AI-driven optimization transforms how cargo is handled and transported across global supply chains. Smart algorithms streamline port operations through automated crane scheduling, berth allocation, and container tracking. AI also supports dynamic route optimization, helping ships identify the most fuel-efficient paths based on sea currents, weather forecasts, and traffic density [32]. These advancements lead to reduced fuel consumption, lower greenhouse gas emissions, and more sustainable maritime practices.

Looking ahead, the integration of AI with autonomous vessels and smart port ecosystems is expected to revolutionize the maritime industry. Autonomous ships equipped with advanced perception and control systems are already being tested to operate with minimal human intervention. Meanwhile, smart ports use AI-powered analytics, robotics, and Internet of Things (IoT) connectivity to improve throughput and enhance security measures. Together, these innovations lay the foundation for a more intelligent, resilient, and environmentally responsible maritime transportation network [33].

3. Conclusion

Artificial intelligence has become a key enabler in modern transportation, offering powerful tools that elevate how mobility systems are managed and experienced [34]. Its integration clearly supports smoother operations, safer travel conditions, and reduced environmental strain across diverse transportation networks. The widespread use of AI—from autonomous driving to predictive analytics—demonstrates its value in improving performance across road, rail, air, maritime, and public transport sectors. By harnessing intelligent algorithms and real-time data processing, transportation agencies can make faster, more accurate decisions that strengthen system reliability and user satisfaction. Ultimately, AI-driven innovations are setting the foundation for a future in which transportation becomes more connected, resilient, and sustainable for communities worldwide.

References

[1] Stutz, Dalmo, Joaquim T. de Assis, Asif A. Laghari, Abdullah A. Khan, Nikolaos Andreopoulos, Andrey Terziev, Anand Deshpande, Dhanashree Kulkarni, and Edwiges GH Grata. "Enhancing security in cloud computing using artificial intelligence (AI)." *Applying Artificial Intelligence in Cybersecurity Analytics and Cyber Threat Detection* (2024): 179–220.

[2] Stutz, Dalmo, Joaquim T. de Assis, Asif A. Laghari, Abdullah A. Khan, Anand Deshpande, Dhanashree Kulkarni, Andrey Terziev, Maria A. de Jesus, and Edwiges GH Grata. "Cyber threat detection and mitigation using artificial intelligence–A cyber-physical perspective." *Applying artificial intelligence in cybersecurity analytics and cyber threat detection* (2024): 107–133.

[3] Kulkarni, D.S. and Rodd, S.F. (2018). Extensive study of text based methods for opinion mining. In *2018 2nd International Conference on Inventive Systems and Control (ICISC)*, 523–527. IEEE.

[4] Kulkarni, D.S. and Rodd, S.S. (2021). Sentiment analysis in Hindi—A survey on the state-of-the-art techniques. *Transactions on Asian and Low-Resource Language Information Processing*, *21*(1), 1–46.

[5] Deshpande, Anand, Vania V. Estrela, and Navid Razmjooy, eds. *Computational intelligence methods for super-resolution in image processing applications*. Cham, Switzerland: Springer International Publishing, 2021.

[6] Deshpande, Anand, Prashant Patavardhan, Vania V. Estrela, and Navid Razmjooy. "Deep learning as an alternative to super-resolution imaging in UAV systems." Imaging and sensing for unmanned aircraft systems 2 (2020): 9.

[7] Iyer, L.S. (2021). AI enabled applications towards intelligent transportation. *Transportation Engineering*, *5*, 100083.

[8] Yan, Huan, and Yong Li. "Generative AI for intelligent transportation systems: Road transportation perspective." *ACM Computing Surveys* 57, no. 12 (2025): 1–45.

[9] Hassan, M., Mahin, H.D., Al Nafees, A., Paul, A. and Shraban, S.S. (2025). Big data applications in intelligent transport systems: A bibliometric analysis and review. *Discover Civil Engineering*, *2*(1), 49.

[10] Haghighat, A.K., Ravichandra-Mouli, V., Chakraborty, P., Esfandiari, Y., Arabi, S. and Sharma, A. (2020). Applications of deep learning in intelligent transportation systems. *Journal of Big Data Analytics in Transportation*, *2*(2), 115–145.

[11] Montoya-Torres, J.R., Moreno, S., Guerrero, W.J. and Mejía, G. (2021). Big data analytics and intelligent transportation systems. *IFAC-PapersOnLine*, *54*(2), 216–220.

[12] Butt, M.S. and Shafique, M.A. (2025). A literature review: AI models for road safety for prediction of crash frequency and severity. *Discover Civil Engineering*, *2*(1), 99.

[13] Tselentis, D.I., Papadimitriou, E. and van Gelder, P. (2023). The usefulness of artificial intelligence for safety assessment of different transport modes. *Accident Analysis & Prevention*, *186*, 107034.

[14] Zhang, N., Tian, R. and Fu, G. (2024). Design and application of automatic driving emergency collision avoidance control algorithm based on artificial intelligence technology. *Measurement: Sensors*, *33*, 101248.

[15] Jevinger, Å., Zhao, C., Persson, J.A. and Davidsson, P. (2024). Artificial intelligence for improving public transport: A mapping study. *Public Transport*, *16*(1), 99–158.

[16] Chen, W., Men, Y., Fuster, N., Osorio, C. and Juan, A.A. (2024). Artificial intelligence in logistics optimization with sustainable criteria: A review. *Sustainability*, *16*(21), 9145.

[17] Gerum, P.C.L., Altay, A. and Baykal-Gürsoy, M. (2019). Data-driven predictive maintenance scheduling policies for railways. *Transportation Research Part C: Emerging Technologies*, *107*, 137–154.

[18] Naeem, M.A., Chaudhary, S. and Meng, Y. (2024). Road to efficiency: V2v enabled intelligent transportation system. *Electronics*, *13*(13), 2673.

[19] Phusakulkajorn, W., Núñez, A., Wang, H., Jamshidi, A., Zoeteman, A., Ripke, B., Dollevoet, R., De Schutter, B. and Li, Z. (2023). Artificial intelligence in railway infrastructure: Current research, challenges, and future opportunities. *Intelligent Transportation Infrastructure*, volume 2, liad016.

[20] Durlik, I., Miller, T., Kostecka, E. and Tuński, T. (2024). Artificial intelligence in maritime transportation: A comprehensive review of safety and risk management applications. *Applied Sciences*, *14*(18), 8420.

[21] Barth, M.J., Wu, G. and Boriboonsomsin, K. (2015). Intelligent transportation systems and greenhouse gas reductions. *Current Sustainable/Renewable Energy Reports*, *2*(3), 90–97.

[22] Rinchi, O., Alsharoa, A., Shatnawi, I. and Arora, A. (2024). The role of intelligent transportation systems and artificial intelligence in energy efficiency and emission reduction. *arXiv preprint arXiv:2401.14560*.

[23] Yang, Z., Zheng, Z., Kim, J. and Rakha, H. (2024). Eco-driving strategies using reinforcement learning for mixed traffic in the vicinity of signalized intersections. *Transportation Research Part C: Emerging Technologies*, *165*, 104683.

[24] Tang, R., De Donato, L., Besinović, N., Flammini, F., Goverde, R.M., Lin, Z., ... and Wang, Z. (2022). A literature review of artificial intelligence applications in railway systems. *Transportation Research Part C: Emerging Technologies*, *140*, 103679.

[25] Li, H., Li, N., Kolmanovsky, I. and Girard, A. (2022). Energy-efficient autonomous driving using cognitive driver behavioral models and reinforcement learning. In *AI-enabled Technologies for Autonomous and Connected Vehicles*, Yi Lu Murphey, Ilya Kolmanovsky, Paul Watta, 283–305. Cham: Springer International Publishing.

[26] Bešinović, N., De Donato, L., Flammini, F., Goverde, R.M., Lin, Z., Liu, R., ... and Vittorini, V. (2021). Artificial intelligence in railway transport: Taxonomy, regulations, and applications. *IEEE Transactions on Intelligent Transportation Systems*, *23*(9), 14011–14024.

[27] Singh, P., Dulebenets, M.A., Pasha, J., Gonzalez, E.D.S., Lau, Y.Y. and Kampmann, R. (2021). Deployment of autonomous trains in rail transportation: Current trends and existing challenges. *IEEE Access*, *9*, 91427–91461.

[28] Brittain, M. and Wei, P. (2019). Autonomous air traffic controller: A deep multi-agent reinforcement learning approach. *arXiv preprint arXiv:1905.01303*.

[29] Durlik, I., Miller, T., Kostecka, E., Łobodzińska, A. and Kostecki, T. (2024). Harnessing AI for sustainable shipping and green ports: Challenges and opportunities. *Applied Sciences*, *14*(14), 5994.

[30] Simion, D., Postolache, F., Fleacă, B. and Fleacă, E. (2024). AI-driven predictive maintenance in modern maritime transport—Enhancing operational efficiency and reliability. *Applied Sciences, 14*(20), 9439.

[31] Durlik, I. et al. (2024). Artificial intelligence in maritime transportation. *Applied Sciences/ Marine Tech Journal.*

[32] Ryan Pannell and J. Mark Munoz (2024). Utilizing AI for maritime transport optimization. Center for Maritime Research Report.

[33] Matthew, N.O. Sadiku, Matthias Oteniya, Janet O. Sadiku (2025). Artificial Intelligence in Maritime Industry In International Journal of Trend in Scientific Research and Development Volume 9 Issue 4, Jul-Aug 2025.

[34] Estrela, V.V., Deshpande, A., Stutz, D., de Assis, J.T., Jumani, A.K., Khan, A.A., ... and Lin, Y.D. (2023). Ad hoc networks in healthcare intelligent transportation systems humanitary relief–MANETs, VANETs, and FANETs. In *Intelligent Healthcare Systems*, 66–89. CRC Press.

2 | Intelligent Transportation Systems: A Comprehensive Review of Technologies, Applications, and Future Trends

Angela Maria Alves[1], Gabriel Gomes de Oliveira[2*], Euclides Lourenco Chuma[3], Gabriel Caumo Vaz[4] and Yuzo Iano[5]

Intelligent transportation systems (ITS) represent a pivotal evolution in addressing the escalating complexities of modern transportation, encompassing challenges such as chronic traffic congestion, increased accident rates, environmental degradation, and the surging demand for efficient mobility of people and goods. This literature review provides a comprehensive examination of ITS, delving into its foundational components, diverse applications across both urban and rural environments, and the pressing challenges and promising future trends that define this rapidly advancing field. We explore key technological enablers including advanced sensing technologies (e.g., inductive loops, radar, LiDAR, video analytics), robust communication networks (e.g., DSRC, 5G, fiber optics), sophisticated control and management systems (e.g., adaptive traffic signals, ramp metering, incident management platforms), and effective information dissemination systems (e.g., Advanced Traveler Information Systems, integrated fare payments). The review highlights the transformative impact of ITS on enhancing safety, improving efficiency, fostering sustainability, and optimizing mobility in various contexts. Furthermore, we address critical challenges such as cybersecurity vulnerabilities, issues of infrastructure interoperability, data privacy concerns, and funding limitations. Looking ahead, the article discusses the profound influence of emerging technologies like artificial intelligence (AI), machine learning (ML), autonomous vehicles (AVs), Vehicle-to-everything (V2X) communication, the Internet of Things (IoT), edge computing, blockchain, and Mobility-as-a-Service (MaaS) on shaping the future of transportation. By synthesizing current knowledge and identifying key areas for future research, this review aims to provide a valuable resource for researchers, policymakers, and practitioners engaged in the development and implementation of next-generation intelligent transportation solutions.

[1] Dom Pedro I Highway (SP-65), Km 143,6 - Chácaras Campos dos Amarais, Campinas - SP, 13069-901, Brazil.
[2,4,5] School of Electrical and Computer Engineering, State University of Campinas, 400 Albert Einstein Ave., Cidade Universitária, Campinas, SP 13083-852, Brazil.
[3] Paschoal Marmo, 1888 - Jardim Nova Itália, Limeira - SP, 13484-332, Brazil.
* Corresponding author: ggomes@unicamp.br

DOI: 10.1201/9781003648147-02

1. Introduction

Transportation systems worldwide face escalating challenges, including chronic traffic congestion, increased accident rates, environmental pollution, and the growing demand for efficient mobility of people and goods. These issues are particularly acute in rapidly urbanizing regions and, concurrently, present unique safety and connectivity challenges in sprawling rural landscapes [1]. Traditional approaches, primarily focused on expanding physical infrastructure (e.g., building new roads, widening existing ones), are often insufficient and unsustainable due to immense financial outlays, significant social disruption, and adverse environmental impacts [2, 3]. The limitations of a purely infrastructure-centric model have necessitated a paradigm shift toward intelligent solutions [4]. Intelligent transportation systems (ITS) emerged as a transformative solution to these pressing challenges. ITS integrates cutting-edge information and communication technologies (ICT) directly into transportation networks and vehicles, creating a dynamic and responsive ecosystem [4]. This integration goes beyond simple automation; it involves the intelligent collection, processing, and dissemination of real-time data to inform decision-making for a diverse range of stakeholders—from individual travelers and fleet operators to infrastructure managers and emergency services [5].

The overarching goal of ITS is multifaceted: to create a safer, more efficient, sustainable, and user-centric transportation environment. By leveraging technology, ITS aims to [6]:

- *Enhance safety*: Reduce traffic accidents and fatalities through proactive warnings, improved incident response, and better situational awareness.
- *Improve efficiency*: Optimize traffic flow, minimize congestion, reduce travel times, and enhance the productivity of freight and public transportation.
- *Foster sustainability*: Decrease fuel consumption, lower greenhouse gas emissions, and promote the use of more environmentally friendly modes of transport.
- *Enhance mobility and accessibility*: Provide reliable and timely travel information, facilitate seamless intermodal transfers, and improve access to transportation for all users.

The conceptual framework of ITS can be visualized as a synergistic interplay of sensing, communication, processing, and application layers, as illustrated in Figure 2.1. Each layer contributes to the holistic functionality of an ITS. This article delves into the key components and technologies underpinning ITS, explores its diverse applications in both urban and rural contexts, and finally discusses the pressing challenges and promising future trends that are shaping the evolution of this critical field [7]. It consolidates existing knowledge and presents a critical analysis that serves as a reference point for researchers, industry professionals, and policymakers interested in the advancement and effective implementation of these transformative solutions.

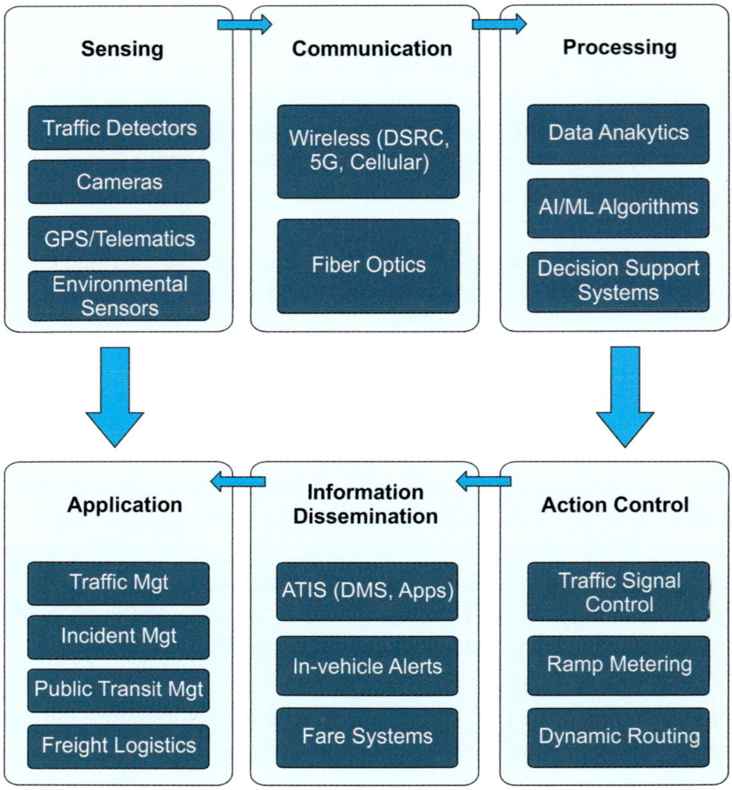

Fig. 2.1 Conceptual framework of intelligent transportation systems (ITS) https://doi.org/10.1080/23311916.2024.2427235

2. Key Components and Technologies of ITS

The effectiveness of ITS hinges on the seamless integration of various technological components that work in concert to gather, analyze, and act upon transportation data. The categorization of these components is discussed in this section [8].

2.1 Data Collection and Sensing Technologies

Sensors are the "eyes and ears" of ITS, providing real-time data on traffic, vehicle movements, and environmental factors [9, 10]. Without this precise and continuous data, effective decision-making would be impossible. Several types of sensors are used, including inductive loop sensors embedded in roads to detect vehicles and measure their speed and occupancy. Radar and LiDAR sensors use radio waves and laser pulses to precisely measure vehicle speed and distance [11].

Video cameras capture visual data at intersections and along highways for traffic monitoring and incident detection, with advanced analytics often powered by artificial intelligence (AI) [12]. Finally, GPS and telematics systems in vehicles provide real-time location and speed data, which is crucial for navigation, fleet management, and traffic monitoring [13]. These components are the backbone of ITS, transforming how we manage and experience mobility.

2.2 Communication Networks

Robust and reliable communication is fundamental for transmitting data between various ITS components, vehicles, and infrastructure. For vehicular communication to be successful, however, innovative infrastructures are essential. For example, communication systems are essential for connecting the various components of an ITS [14]. Dedicated Short Range Communication (DSRC) is a wireless technology specifically for vehicle-to-vehicle (V2V) and vehicle-to-infrastructure (V2I) communication, which enables features like collision warnings. The introduction of 5G networks offers high-speed, low-latency connectivity, supporting advanced applications such as autonomous vehicles (AVs) and real-time data analytics [15]. Fiber optic networks provide high-capacity, low-latency connections, especially for backhaul and data centers. Additionally, existing cellular networks like GSM are widely used to transmit data from vehicles and roadside units to central management systems, particularly where dedicated ITS infrastructure is limited [16].

2.3 Control and Management Systems

These systems process the collected data and implement strategies to optimize transportation operations. ITS offers diverse solutions to enhance urban and road mobility.

Operational management systems are key to ITS [11]. Adaptive traffic signal control systems adjust signal timings based on real-time traffic to reduce wait times and improve flow at intersections. Ramp metering systems control the rate at which vehicles enter highways to prevent congestion. Incident management systems use real-time sensor and camera data for rapid detection of and response to traffic incidents [17]. Public Transportation Management Systems (PTMS) optimize public transport with features like real-time vehicle tracking and dynamic scheduling. Finally, Freight and Commercial Vehicle Operations (CVO) systems provide real-time tracking, route optimization, and predictive maintenance to improve the efficiency of logistics fleets [18].

2.4 Information Dissemination Systems

Advanced Traveler Information Systems (ATIS) provide drivers with pre-trip and en route information through various channels like dynamic message signs, smartphone apps, and in-vehicle navigation [17]. This information includes traffic congestion, road closures, and alternative routes. Integrated fare payment systems streamline public transport by offering convenient, unified payment options that work across different modes of transportation [11].

A fundamental pillar of ITS is providing accurate and real-time information to travelers, empowering them to make well-informed decisions. This is made possible by solutions like ATIS, which offer data on routes, traffic conditions, and alternative transportation options. Additionally, Integrated Fare Payment Systems simplify access to different modes of transport, removing barriers and making mobility more fluid and efficient for everyone.

3. Applications of ITS in Urban and Rural Areas

The ITS applications are diverse and tailored to address the specific challenges and opportunities presented by both urban and rural environments.

3.1 Urban Applications

Urban areas, characterized by high population density and chronic congestion, have been the main focus of early ITS implementations. Given this, management focused on optimizing time, reducing costs, and lessening environmental impact has been adopted in large cities through various actions.

Intelligent transportation systems address several key areas of modern transport. For congestion management, they use adaptive traffic signals, ramp metering, and dynamic rerouting to reduce travel times and improve traffic flow [11]. To optimize public transport, ITS provides real-time tracking, electronic fare collection, and passenger information, making it more efficient and appealing. Parking management is improved through smart systems that guide drivers to open spots, reducing search time and congestion [19]. For emergency management, ITS provides real-time incident data and optimized routes to help first responders arrive faster. Finally, by optimizing traffic flow and reducing idling, ITS helps lower fuel consumption and greenhouse gas emissions, thereby contributing to pollution reduction [11, 15].

3.2 Rural Applications

While rural areas may experience less frequent congestion, they face unique challenges such as long distances, limited infrastructure, and higher rates of severe crashes.

Intelligent transportation systems offer numerous benefits for rural areas, enhancing safety and efficiency [20]. Improved road safety is a primary advantage, with real-time monitoring and alerts for hazardous conditions like icy roads and wildlife, which can reduce crashes. Incident management and response are also significantly improved, as ITS can quickly identify incident locations and provide accurate information to emergency services, which is crucial in areas where they might be far away [21]. Traveler information services are vital in rural areas with dispersed services, offering real-time information on road conditions, weather, and service availability. Connected vehicle technologies, such as V2V and V2I communication, can improve collision avoidance at rural intersections and blind curves. Additionally, ITS can be used for the remote monitoring of infrastructure like bridges and roads, enabling proactive maintenance. For commercial operations, ITS helps optimize routes for agricultural and delivery vehicles, leading to fuel savings and greater efficiency [16, 22].

Intelligent transportation systems are crucial for improving mobility in both urban and rural environments. They use sensors and advanced technologies to optimize traffic flow, manage incidents, and enhance safety and efficiency for vehicles and public transit. Essentially, ITS transforms transportation with real-time data for intelligent decision-making, leading to safer, faster, and more sustainable journeys.

4. Discussion

Intelligent transportation systems are changing how we get around, offering powerful ways to tackle the tough problems in today's transportation. We are talking about everything from frustrating traffic jams and increasing accident rates to the environmental impact of our daily commutes and the constant need for smooth, easy travel. ITS is key to solving these issues, completely transforming our cities and rural areas.

At its core, ITS uses a clever combination of different technologies. Imagine it as a highly connected "nervous system" for our roads and transit. Advanced sensors—things like inductive loops, radar, LiDAR, and video analysis—act as the system's eyes and ears. They gather real-time information on traffic flow, incidents, and road conditions. This data then travels quickly and reliably through strong communication networks, including DSRC, 5G, and fiber optics.

But collecting data is just the beginning. Sophisticated control and management systems are the brains of ITS. They use this information to make smart decisions, like adjusting traffic lights in real time, managing highway access with ramp meters, and quickly responding to accidents and breakdowns through incident management platforms. Finally, effective information sharing systems, such as ATIS and integrated fare payments, keep us, the travelers, informed and our trips running smoothly.

By understanding what makes up ITS, how it is used, the challenges it faces, and where it is headed, we can all work together to build transportation systems that are smarter, safer, and more sustainable for everyone.

4.1 Future Trends and Innovations

The future of ITS is being shaped by several cutting-edge technologies and evolving paradigms, promising even more intelligent, connected, and autonomous transportation systems. These trends are not isolated but rather interconnected, fostering a holistic transformation of how we move people and goods. The main trends in technology use within the mobility segment are discussed in this section.

4.1.1 Artificial Intelligence and Machine Learning

Artificial intelligence (AI) and machine learning (ML) are increasingly being integrated into ITS for predictive analytics, real-time decision-making, anomaly detection, and optimization of complex traffic scenarios. This includes AI-powered traffic prediction models that can anticipate congestion before it occurs, adaptive signal control systems that learn and optimize timings based on real-time and historical data, and intelligent incident detection algorithms that can identify and classify events with greater accuracy [23]. Furthermore, ML can enhance the capabilities of AVs by improving their perception, decision-making, and planning algorithms.

4.1.2 Autonomous Vehicles and Connected and Automated Vehicles

The development and deployment of AVs represent a transformative trend. While fully autonomous vehicles are still evolving, connected and automated vehicle (CAV) technology, which allows vehicles to communicate with each other (V2V) and with infrastructure (V2I), holds immense potential for improving traffic flow, enhancing safety through collision avoidance, and optimizing vehicle efficiency, particularly in applications like truck platooning [24]. CAVs are expected to significantly reduce human error, which is a leading cause of accidents.

4.1.3 Vehicle-to-Everything (V2X) Communication

Vehicle-to-everything (V2X) is an umbrella term encompassing V2V, V2I, vehicle-to-pedestrian (V2P), and vehicle-to-network (V2N) communication. It is expected to play a crucial role in improving safety, traffic management, and driving efficiency by enabling vehicles to exchange information with their surroundings [25, 26]. For example, V2P could alert drivers to pedestrians in blind spots, while V2N could provide real-time traffic updates from a centralized cloud. The advent of 5G connectivity is a key enabler for the low-latency and high-bandwidth requirements of V2X.

4.1.4 Internet of Things (IoT) Integration

The proliferation of Internet of Things (IoT) devices will lead to an even richer stream of data from vehicles, infrastructure (smart traffic lights, smart parking sensors), and even individuals (wearable devices). This extensive network of interconnected sensors will enable more granular and real-time insights into transportation dynamics, allowing for highly responsive and personalized transportation services [27].

4.1.5 Edge Computing

Processing data closer to the source (at the "edge" of the network), rather than sending it to a central cloud, can significantly reduce latency. This is critical for real-time ITS applications like autonomous driving (where instantaneous decision-making is crucial) and rapid incident response, as it minimizes the delay in data transmission and processing [28].

4.1.6 Blockchain Technology

Blockchain could offer secure and transparent ways to manage data in ITS, particularly for secure data sharing between disparate entities, facilitating reliable payment systems for mobility services, and verifying the authenticity of information in connected environments. This can enhance trust and data integrity within the ITS ecosystem [29].

4.1.7 Mobility-as-a-Service (MaaS)

Mobility-as-a-Service (MaaS) platforms integrate various transportation services (public transit, ride-sharing, bike-sharing, car-sharing, autonomous shuttles) into a single, on-demand service. These platforms often leverage ITS data to provide seamless and efficient multimodal travel options, allowing users to plan, book, and pay for their journeys through a single application. This represents a significant shift from private vehicle ownership to a more service-oriented and sustainable approach to mobility.

5. Conclusion

Intelligent transportation systems represent a vital evolution in addressing the complex challenges of modern mobility. By integrating advanced sensing, communication, and control technologies, ITS offers a comprehensive approach to enhancing safety, improving efficiency, reducing environmental impact, and providing better traveler information. From urban congestion management to improving safety and connectivity in rural areas, the applications of ITS are vast and continuously expanding.

While significant challenges remain, particularly in cybersecurity, interoperability, and funding, the ongoing advancements in AI, AVs, V2X communication, and IoT promise a future where transportation systems are not only intelligent but also highly adaptive, resilient, and responsive to the needs of society [30].

Continued research, collaboration among stakeholders, and strategic investments are essential to fully realize the transformative potential of ITS and to pave the way for a truly smart and sustainable transportation future.

Finally, future ITS efforts must prioritize continued research and strategic investments in AI, AVs, and V2X communication, while diligently addressing cybersecurity, interoperability, and funding challenges. Collaborative stakeholder engagement is crucial to fully realize the potential of ITS, fostering adaptive, resilient, and responsive transportation systems for a smarter, sustainable future of mobility.

References

[1] Kinkus, J. (2009). Alternative fuels and advanced vehicles data center. *Reference Reviews*, *23*(5), 42–43.

[2] Chowdhury, M. and Dey, K. (2016). Intelligent transportation systems—A frontier for breaking boundaries of traditional academic engineering disciplines [education]. *IEEE Intelligent Transportation Systems Magazine*, *8*(1), 4–8.

[3] Moradi, M. and Assaf, G.J. (2023). Designing and building an intelligent pavement management system for urban road networks. *Sustainability*, *15*(2), 1157.

[4] Knaian, A.N. (2000). A wireless sensor network for smart roadbeds and intelligent transportation systems (Unpublished doctoral dissertation). Massachusetts Institute of Technology.

[5] Lu, C. and Wang, F. (2024). Towards secure Internet of Things-enabled intelligent transportation systems: A comprehensive review. *International Journal of Advanced Computer Science & Applications*, *15*(7). The requested information is not available: https://dx.doi.org/10.14569/IJACSA.2024.0150708

[6] Avcı, I. and Koca, M. (2024). Intelligent transportation system technologies, challenges and security. *Applied Sciences*, *14*(11), 4646.

[7] Rammohan, A., et al. (2023). Revolutionizing intelligent transportation systems with cellular vehicle-to-everything (C-V2X) technology: Current trends, use cases, emerging technologies, standardization bodies, industry analytics and future directions. *Vehicular Communications*, *43*, 100638.

[8] Amato, F., Cozzolino, G., Moscato, V. and Moscato, F. (2019). Analyse digital forensic evidences through a semantic-based methodology and NLP techniques. *Future Generation Computer Systems*, *98*, 297–307.

[9] Creß, C., Bing, Z. and Knoll, A.C. (2023). Intelligent transportation systems using roadside infrastructure: A literature survey. *IEEE Transactions on Intelligent Transportation Systems*, *25*(7), 6309–6327.

[10] Singh, B. and Gupta, A. (2015). Recent trends in intelligent transportation systems: A review. *Journal of Transport Literature*, *9*(2), 30–34.

[11] Torre-Bastida, A.I., Del Ser, J., Laña, I., Ilardia, M., Bilbao, M.N. and Campos-Cordobés, S. (2018). Big data for transportation and mobility: Recent advances, trends and challenges. *IET Intelligent Transport Systems*, *12*(8), 742–755.

[12] Wang, W., Krishnan, R. and Diehl, A. (2015). Advances and challenges in intelligent transportation: The evolution of ICT to address transport challenges in developing countries. *Transport and ICT Connections*, *26*. The requested information is not available: https://documents1.worldbank.org/curated/en/408361638167733733/pdf/Advances-and-Challenges-in-Intelligent-Transportation-The-Evolution-of-ICT-to-Address-Transport-Challenges-in-Developing-Countries.pdf

[13] Kumar, M., Albert, S. and Deeter, D. (2005). A summary of rural intelligent transportation systems (ITS) benefits as applied to ODOT region 1. Oregon Department of Transportation Region, 1.

[14] Rahill, M. (2020). Alternative fuels data center and related technology integration tools (Tech. Rep.). National Renewable Energy Lab. (NREL), Golden, CO (United States).

[15] Ebrahimpour, A., Baibordy, A. and Ibrahim, A. (2025). Developing a research roadmap for highway bridge infrastructure innovation: A case study. *Infrastructures*, *10*(6), 133.

[16] Daghigh, R., Pishvaee, M. S., Jabalameli, M. S. and Pakseresht, S. (2022). A possibilistic–stochastic programming approach to resilient natural gas transmission network design problem under disruption: A case study. *Journal of Industrial and Systems Engineering*, *14*(1), 137–162.
[17] Stehlin, J. and Teron, L. (2025). States of repair: The role(s) of state government in shaping urban highway removal in New York state. *Journal of Urban Affairs*, 1–29. The requested information is not available: https://doi.org/10.1080/07352166.2025.2473954
[18] Rostami, M. and Goli-Bidgoli, S. (2024). An overview of QoS-aware load balancing techniques in SDN-based IoT networks. *Journal of Cloud Computing*, *13*(1), 89.
[19] Veres, M. and Moussa, M. (2019). Deep learning for intelligent transportation systems: A survey of emerging trends. *IEEE Transactions on Intelligent Transportation Systems*, *21*(8), 3152–3168.
[20] Mfenjou, M.L., Ari, A.A.A., Abdou, W., Spies, F. et al. (2018). Methodology and trends for an intelligent transport system in developing countries. *Sustainable Computing: Informatics and Systems*, *19*, 96–111.
[21] Beck, D., Ferasso, M., Storopoli, J. and Vigoda-Gadot, E. (2023). Achieving the sustainable development goals through stakeholder value creation: Building up smart sustainable cities and communities. *Journal of Cleaner Production*, *399*, 136501.
[22] Maynor, D. (2011). *Metasploit toolkit for penetration testing, exploit development, and vulnerability research*. Elsevier.
[23] Chen, Y., Zhan, X., Yang, W., Yan, X., Du, Y. and Zhao, T. (2025). A NLP analysis of digital demand for healthcare jobs in China. *Scientific Reports*, *15*(1), 14518.
[24] Shen, T., Hua, K. and Liu, J. (2019). Optimized public parking location modelling for green intelligent transportation system using genetic algorithms. *IEEE Access*, *7*, 176870–176883.
[25] Jaouadi, M.H.O. (2022). Investigating the influence of big data analytics capabilities and human resource factors in achieving supply chain innovativeness. *Computers & Industrial Engineering*, *168*, 108055.
[26] Wang, F.-Y., Mirchandani, P. and Zheng, N. (2004). Advances and trends in research and development of intelligent transportation systems: An introduction to the special issue. *IEEE Transactions on Intelligent Transportation Systems*, *5*(4), 222–223.
[27] Zhu, S., Yu, T., Xu, T., Chen, H., Dustdar, S., Gigan, S., . . . et al. (2023). Intelligent computing: the latest advances, challenges, and future. *Intelligent Computing*, *2*, 0006.
[28] Ge, Y., Zhang, G., Meqdad, M. N. and Chen, S. (2023). A systematic and comprehensive review and investigation of intelligent IoT-based healthcare systems in rural societies and governments. *Artificial Intelligence in Medicine*, *146*, 102702.
[29] Gong, T., Zhu, L., Yu, F.R. and Tang, T. (2023). Edge intelligence in intelligent transportation systems: A survey. *IEEE Transactions on Intelligent Transportation Systems*, *24*(9), 8919–8944.
[30] Stutz, D., de Assis, J.T., Laghari, A.A., Khan, A.A., Deshpande, A., Kulkarni, D., ... and Grata, E. G. (2024). Cyber threat detection and mitigation using artificial intelligence—A cyber-physical perspective. In *Applying Artificial Intelligence in Cybersecurity Analytics and Cyber Threat Detection*, eds S, Mahajan, M. Khurana, & V. V. Estrela, 107–133. Wiley.

3 Role of AI in Traffic Safety and Prevention of Accidents

Gujanatti Rudrappa[1*], Arun Tigadi[2], Sushant Jadhav[3], Kartik Cholachgudda[4], Ravi Shankar D.[5], Bhanuteja G.[6] and Sachin Urabinahatti[7]

Artificial intelligence (AI) will reduce accident risk and allow vehicles to move safely, making it the new face of automotive safety and accident avoidance. This chapter discusses AI applications in real-time traffic monitoring, predictive accident risk assessment, and autonomous driving. An AI algorithm utilizes machine learning and deep learning to identify high-risk events in traffic camera, sensor, and vehicular communication channel data, advising drivers and traffic control systems. AI has also used computer vision and sensor fusion–based Advanced Driver-Assistance Systems (ADAS) to reduce traffic mishaps, prevent crashes, and enable emergency autonomous maneuvering. Accidents are avoided using predictive maintenance models for vehicles and buildings. Real-time signal light timing and re-route tactics lessen traffic congestion and accidents. AI could remove human errors that cause most road accidents. AI's ethical and legal issues require public trust and safety. This study assesses existing studies, identifies challenges, and proposes ways to enhance traffic safety by leveraging AI to improve transportation system safety, intelligence, and resilience.

1. Introduction

1.1 Background and Significance of Traffic Safety

The World Health Organization (WHO) reports that traffic accidents cause 1.35 million deaths and more than 50 million injuries each year worldwide [1]. Human error remains the leading cause of accidents, accounting for more than 90% of incidents, despite improvements in vehicle safety systems/road infrastructure and public awareness programs [2]. High-risk situations in traffic result from driver distractions combined with excessive speeds, driver tiredness, and

[1-3] Department of Electronics and Communication Engineering, KLE Technological University Dr. M. S. Sheshgiri Campus, Belagavi, Karnataka, India.

[4-6] School of Electronics and Communication Engineering, REVA University, Bengaluru, Karnataka, India.

[7] Department of Computer Science Engieering (Artificial Intelligence and Machine Learning) Sai Vidya Institute of Technology, Bangalore, Karnataka, India.

* Corresponding author: rudraguj@gmail.com

DOI: 10.1201/9781003648147-03

impaired decision-making abilities. The rising urban population, combined with the increasing number of vehicles, has made traffic management (TM) safety a crucial issue that transportation authorities, along with policymakers, must address. The implementation of artificial intelligence (AI) in traffic safety systems holds promise for addressing safety challenges by reducing human errors and identifying hazardous points, as well as automating critical system decisions [3]. Real-time data analysis, combined with machine learning (ML) algorithms and advanced sensing technologies, enables AI systems (AISs) to detect dangerous situations more quickly than human operators, leading to improved traffic safety measures [4, 5, 6].

Originally, AI transportation applications began to emerge in the 1980s, when expert systems were used to optimize traffic signals and routes [7]. The advancements in ML, computer vision (CV), and data analytics over the past decades have motivated the development of intelligent transportation systems (ITS) that utilize AI to predict traffic, manage incidents, and make fully automated decisions [8]. The self-driving industry has become a reality, and corporations like Tesla, Waymo, and NVIDIA apply AI to build self-driving systems with the ability to prevent accidents and operate independently [9]. Traffic control with the help of AI is not limited to self-driving cars, but also help identify traffic violation, achieve communication between cars via IoT, etc. which has the capabilities of controlling traffic cameras and sensor networks. These processes provide real-time data to enable continuous learning and optimization of the system [10]. This conversion of conventional TM systems (TMSs) into dynamic, adaptable, and data-oriented systems relies on AI, enabling cities to claim their status of smartness.

In this chapter, the authors examine how AI can be utilized to monitor transportation hazards, predict risks, and mitigate dangers. The form is evaluated in three critical areas in the chapter:

- AI-based traffic surveillance and management systems can improve traffic safety through real-time traffic monitoring and intelligent control systems. Advanced Driver Assistance Systems (ADAS) and AI-powered autonomous transportation systems are collaborating to prevent collisions [11, 12, 13] Click or tap here to enter text.
- AI models enable predicting malfunctions of vehicles thereby preventing the accidents and provide safety precautions.
- The use of AI technology allows not only the dispatching of the emergency services but also the investigation of the accident.

Artificial intelligence in traffic safety systems raises ethical, legal, and societal issues that emerge as AI becomes an integral part of existing transportation systems. The chapter also examines the future trends that promise to yield improved traffic safety performance and a reduced rate of accidents. Technological advancements in the domain of AI for traffic safety and accident prevention involve the collaborative efforts of technology developers and transportation authorities, aiming to achieve the common goals of safety and reliability, while fostering community trust.

2. Understanding AI in Traffic Safety

Figure 3.1 presents an overview of the hierarchical structure of various AI technologies employed in traffic safety.

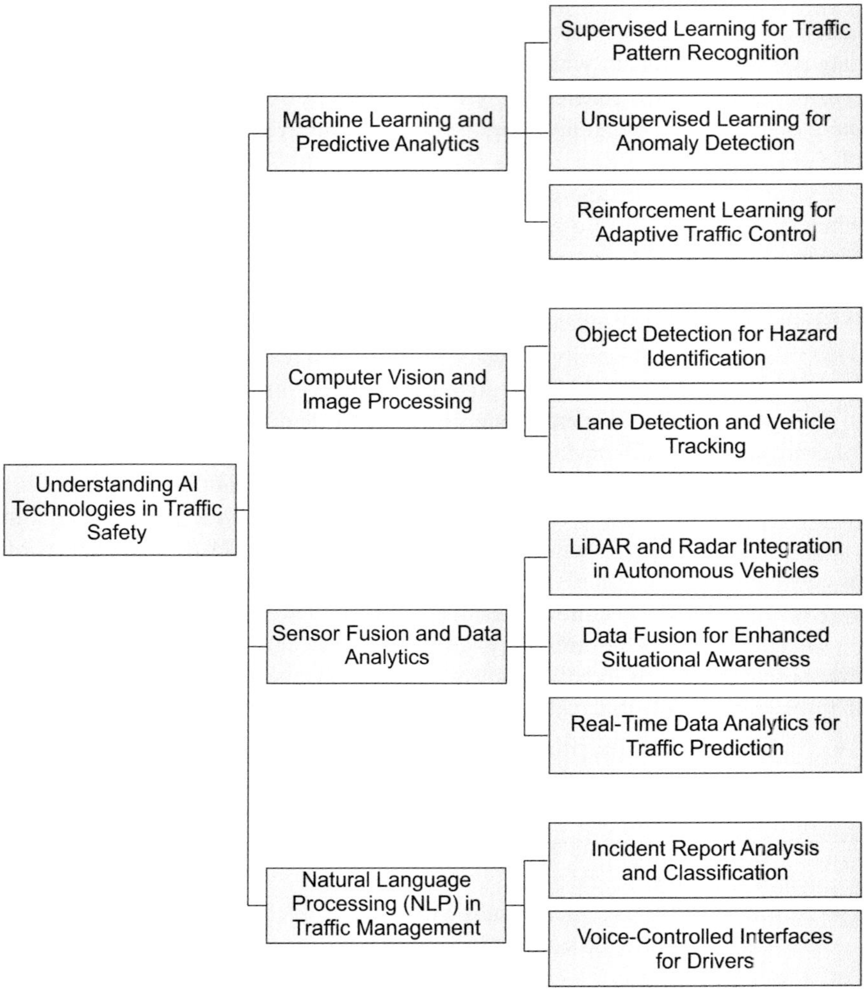

Fig. 3.1 Overview of the hierarchical structure of various AI technologies employed in traffic safety

2.1 Machine Learning and Predictive Analytics

The ML algorithms perfect traffic control systems by processing huge amounts of data relayed by surveillance cameras and sensors, as well as GPS-based devices, to detect risky scenarios.

- *Supervised Learning for Traffic Pattern Recognition*: General traffic trends and hazardous driving activities are outlined all over by supervised learning mechanisms comprising decision trees along with support vector machines (SVMs) and neural networks [8]. The models developed make the required prediction regarding the probability of accidents based on input data indicators encompassing vehicle speed, vehicle distance, and road condition records.
- *Unsupervised Learning for Anomaly Detection*: Unsupervised learning models such as the k-means clustering and autoencoders assist in the discovery of abnormal traffic patterns that could either indicate risky driver behavior or unusual traffic conditions. The models analyze a relatively large range of data to identify deviations from usual traffic trends, allowing officials to identify and avoid hazardous conditions before accidents occur [14].
- *Reinforcement Learning (RL) for Adaptive Traffic Control*: The RL can react more easily to traffic signal-timing provisions, redirect traffic to minimize congestion, and provide resources in response to emergencies [15].

2.2 Computer Vision and Image Processing

Computer vision is considered one of the components of traffic safety since it enables the AISs to receive and process the visual data retrieved via the traffic cameras, dashcams, and car-mounted sensors. Computer Vision (CV)-based procedures analyze video steams and images to identify objects in real time, classify them, detect lane markings, as well as evaluate driving behavior.

The area of traffic safety widely uses YOLO and Faster R-CNN (Regional CNN) object detectors with the purpose of identifying vehicles, pedestrians, cyclists, and road signs [16].

The ADAS and autonomous vehicles use lane detection as the primary component. Hough Transform and Convolutional Neural Network (CNN)–based models of Lane detection cameras embedded in the CV framework check the line markings on the road utilizing the image/video processing techniques to ensure discipline and sound an alert in case of lane departure [17, 18].

2.3 Sensor Fusion and Data Analytics

Sensor fusion will be used to process data measured by Light Detection and Ranging (LiDAR) sensors, radar cameras, and GPS devices, thereby creating a single, comprehensive image of the environment. The fusion of sensor information enhances AIS efficiency in identifying and managing traffic conditions.

The LiDAR creates a detailed three-dimensional map of the environment, while radar identifies objects and determines their speed and distance from the sensor. A combination of these technologies allows better perception and reaction by autonomous vehicles, which is why risks of collisions are reduced to a minimum.

As interpreters, AI systems with sensor fusion methods like Bayesian networks and Kalman filters can integrate data from different sources. This strategy of merging sensors improves situational awareness by overcoming sensor deficiencies and improving traffic understanding.

An AI big-traffic data analytics platform will instantaneously analyze such data to identify patterns that can help it to predict traffic congestion spots, as well as accident-prone areas and possible bottlenecks. The traffic authorities are applying such observations to put in place preventive strategies that reduce the possibility of traffic accidents [19].

2.4 Natural Language Processing in Traffic Management

Natural language processing (NLP) has gained wide application in analyzing textual information regarding traffic incidents, as well as in analyzing remarks and other related messages from drivers.

Accident reports, emergency driver communication, and driver complaints handled using NLP models yield useful data. Text classification algorithms like Bidirectional Encoder Representations from Transformers (BERT) can classify situations, uncover root causes, and choose the most important reaction [20].

The AI voice assistants (utilizing NLP technology) enable navigation and the management of scenic information without requiring hands-on interaction, allowing the driver to maintain complete road attention. The protection level will also be enhanced when drivers operate on voice-controlled interfaces, as these platforms minimize disturbances by providing important instructions in emergencies [21].

3. AI-Enabled Traffic Monitoring and Management

Road accidents and traffic congestion are major issues in cities due to high population density and rising vehicle numbers. Traditional traffic controllers, which use set timing and limited real-time information, may struggle to adapt to traffic dynamics. Live data from sensors, security cameras, and linked vehicles is used by AI-enabled TM systems to make data-driven decisions that enhance traffic situation and reduce congestion [22]. AI is used in traffic monitoring for traffic signal control, predictive traffic analytics, congestion management, and intelligent traffic control in smart cities.

3.1 AI Traffic Monitoring

Monitoring traffic and detection of various behaviors like over-speeding and traffic law violations are necessary in traffic surveillance systems.

Traffic surveillance AI solutions depend on CV and deep learning (DL) algorithms to observe data from video cameras and generate real-time traffic updates. Such systems can identify vehicles, humans, and other objects, group classification, and track movement to estimate vehicular density and pinpoint anomalous behaviors. To detect vehicles and people in complex traffic conditions, standard object detection algorithms can be utilized, such as YOLO and Faster R-CNN. The capabilities of these systems are locating the hotspots of congestion based on the density of the traffic and vehicular speed, indicators of violations, unlawful change of lanes, driving too fast, and automatic activation of alarm to police departments.

Artificial intelligence utilizes sensory data and visual analytics to identify traffic issues, including accidents, vehicle breakdowns, and traffic bottlenecks. The technologies can detect anomalous traffic and trigger alerts, allowing emergency services to respond quickly. AI helps to reduce accidents and improve road safety by minimizing the time between incidents and interventions [23].

The Optical Character Recognition process and DL models that are based on AI usage allow reading and analysis of license plate information by Automatic License Plate Recognition systems. These systems enhance law enforcement control by recognizing vehicles with traffic violations, stolen vehicles, and those involved in criminal activity [24].

3.2 AI in Traffic Signal Control and Optimization

Conventional traffic light systems work according to laid-down timetables, and this could result in poor TM and overcrowding. Adaptive traffic signal control systems (ATSCs) under AI control are capable of dynamic traffic signal timing adjustments according to real-time traffic situations. Such systems incorporate minimum waiting time and congestion reduction of traffic light timings [8].

The AI-reliant TM system can manage traffic at several intersections to maximize the fluidity of traffic in the whole city. The capacities of AI models to estimate congestion patterns based on historical and real-time traffic data enable the adjustment of signal timing. AI-enabled traffic control measures tend to be simulated and assessed in traffic flow optimization models to include Simulation of Urban Mobility and VISSIM [25] [26].

Traffic signal systems with AI also prioritize the safety of pedestrians and cyclists, as an AI-based system can detect the presence of pedestrians or cyclists at an intersection and adjust the signal cycles to ensure safe crossing. LiDAR sensors and image processing help detect vulnerable road users and adjust traffic control mechanisms.

3.3 Predictive Traffic Analytics and Congestion Management

The ML models using predictive analytics can predict traffic congestion using past traffic records and weather conditions. Such time series models as Long Short-Term Memory (LSTM) networks and Auto Regressive Integrated Moving Average models are usually used to predict the future traffic conditions and congestion points. The advantages of such systems include active traffic control through diverting cars to alternative routes, fewer traffic jams, and resultant environmental effects.

The AI-based route optimization systems offer the best routes that drivers should follow based on the current traffic situation and the expected congestion. Route updating and minimization of travel time are achieved by hinging on graph theory and shortest path algorithm, like Dijkstra's algorithm search in these systems [27]. For example, Google Maps and Waze use AI for recommending the best route to be taken and providing real-time traffic information. AI traffic re-routing solutions are being utilized in smart cities to manage large-scale events and alleviate gridlock.

3.4 Intelligent Traffic Control in Smart Cities

The AISs are connected with the devices developed within the Internet of Things (IoT), built to operate with smart cameras, traffic and vehicle (connected) sensors, and smart roads and vehicles in the development of smart cities. AI models process real-time data provided by IoT devices to enhance traffic control, incident management, and emergency response. The AISs capabilities include constant supervision of the traffic situation and environmental effects, and real-time integration of the transport infrastructure and emergency services. Vehicle-to-infrastructure (V2I) communication, in which the vehicles communicate with the traffic control systems and roadside units (RSUs), is enabled by AI. V2I communication improves the awareness of situations, enhances the movement of traffic, as well as guarantees immediate response to possible accidents [28].

4. AI for Accident Prevention and Collision Avoidance

Road accidents are one of the leading causes of death and injury worldwide, killing 1.35 million people yearly, according to the WHO [1]. Traditional accident prevention strategies, which rely on traffic enforcement and post-accident investigation, are often reactive and ineffective. AI provides proactive solutions, leveraging real-time data, predictive analytics, and intelligent decision-making to prevent crashes. It can analyze driver behavior, identify dangers, predict collisions, and dprevent accidents with automated control systems [29]. Self-driving braking systems, ADAS, and AI-controlled collision avoidance are being increasingly utilized to enhance road safety. These systems use CV, sensor fusion, and ML to detect and respond to threats in real time. AI technologies improve driver support and prepare for fully driverless vehicles, which will make transportation safer [30, 31].

4.1 AI Technologies for Collision Avoidance and Risk Mitigation

The CV systems play an important role in collision avoidance since they detect and categorize obstacles, pedestrians, and other vehicles. These systems detect likely hazards with specificity using cameras that take high-resolution images [32, 33] and are installed on vehicles, as well as the latest DL frameworks, such as CNNs and generative adversarial networks (GANs) [34].

Computer vision can be supplemented with LiDAR and radar distance measurement and 3D object mapping systems. LiDAR utilizes laser beam pulses to generate a detailed 3-D profile of the surroundings, enabling it to accurately detect nearby objects. Radar systems sense distance of an object and its velocity, and hence play a key role in adaptive cruise control and collision warning [35].

4.2 AI-enabled Driver Monitoring and Behavior Prediction

The AISs can identify driver inattention due to drowsiness and fatigue through facial expression, eye movement, and the slope of the head. ML models may define facial landmarks and physiology to issue real-time warnings that prevent accidents caused by the driver's inattention [36].

It is possible to define potential collision scenarios using predictive AI models, driver behavior, and the probability of risky moves. RL algorithms and recognition of behavioral patterns are applied to predict lane departures, abrupt braking, and other moves leading to accidents [37].

4.3 Autonomous Emergency Braking (AEB) and Collision Avoidance Systems

The AI-powered autonomous emergency braking (AEB) devices detect potential crashes and automatically brake if the driver does not. This system utilizes sensor fusion to integrate data from cameras, LiDAR, radar, and ultrasonic sensors, thereby understanding the vehicle's surroundings [38].

The AI-driven collision avoidance systems initiate emergency braking and evasive actions when braking fails, analyzing escape movements and selecting the most effective ones using predicted trajectory analysis [39].

4.4 Vehicle-to-Everything (V2X) Communication for Accident Prevention

Vehicle-to-vehicle (V2V) and V2I communication allow vehicles to exchange real-time information regarding their location, velocity, and path. AISs process this information to predict possible collision scenarios and warn drivers or automatically take control [40].

Vehicle-to-pedestrian (V2P) communication enhances pedestrian safety by enabling vehicles to sense and communicate with pedestrians using mobile devices or smart wearables. AISs analyze data from the devices to forecast pedestrian movement and avoid collisions [41].

4.5 AI-powered Decision-making in Autonomous Vehicles

Perception systems in self-driving cars employ AI models to analyze sensory data and generate an overall impression of the driving scene. Perception systems combine information from cameras, LiDAR, radar, and ultrasonic sensors to identify road features and moving objects [42].

Urban electric vehicles (UEVs): Using road geometry/traffic and the location of obstacles, AI in autonomous cars computes safe and efficient trajectories. Predictive models are used to estimate the best possible acceleration, braking, and steering in motion control systems to avoid collisions [43] using algorithms such as Dijkstra's, Kalman, and LSTM.

5. Predictive Maintenance and Infrastructure Safety

Transport infrastructure includes roads, bridges, tunnels, and railroads, which ensure mobility and safety. The gradual degradation caused by aging, traffic loading, and environmental factors creates safety issues. Conventional maintenance relies on routine inspections and reactive repairs to find problems after they become significant. In contrast, AI-based predictive maintenance uses real-time monitoring, sensor-based data analysis, and ML to predict potential malfunctions, reducing safety risks and extending infrastructure system life [44]. AI technologies analyze abnormalities, patterns in prior data, sensor data, and environmental conditions to determine when to intervene and maintain systems and infrastructure. AI-based predictive maintenance optimizes resource utilization, reduces operational costs, and enhances key infrastructure [45] [46]. IoT sensors and cloud analytics provide infrastructure operators with real-time data to make informed decisions and prevent catastrophic failures [47].

5.1 AI Technologies for Predictive Maintenance

Machine learning can aid in anomaly detection in infrastructure systems by analyzing sensor data and identifying abnormal behavior, which serve as indicators of potential system failures. Such models use supervised, unsupervised, and reinforcement learning techniques to extract patterns and can make predictions of failure cases [48, 49] Click or tap here to enter text.. The models used are support vector machines, random forest, decision trees, neural networks, and many more.

Microsoft SHM involves continuous monitoring of the infrastructure's state using sensors, accelerometers, and other IoT devices. DL-based models like CNNs and Recurrent Neural Networks (RNNs) process information from these devices to identify cracks, deformations, and vibrations in real time [50]. The applications include monitoring bridges, detection of micro-cracks, and analysis of seismic activity.

Sensor fusion is the merging of information from several sources, including LiDAR, thermal sensors, and vibration sensors, to enhance predictive performance. IoT sensors mounted on infrastructure transmit data automatically and in real time to cloud platforms, where AI models process and analyze the data to identify anomalies in normal behavior [51]. The capabilities are real-time evaluation of the condition of the road surface and identification of temperature variations that impact asphalt quality.

5.2 AI-powered Risk Assessment and Failure Prediction

Predictive analytics driven by AI enables early risk assessment by identifying high-risk areas and potential failure points within infrastructure systems. Predictive models can be used to prioritize maintenance operations by analyzing past failure history, environmental factors, and usage patterns [52].

Virtual digital twin (DT) technology replicates physical infrastructure, enabling AI models to simulate real-time scenarios and predict outcomes. The models utilize field sensor data to examine structural behavior, stress levels, and environmental variables, thereby predicting failures [53]. Examples include simulation of bridge load conditions, simulation of wear and tear on highways due to heavy traffic, prediction of tunnel degradation considering environmental conditions.

5.3 AI for Traffic Flow Optimization and Infrastructure Longevity

The AI algorithms examine the patterns of traffic to forecast peak loads and to dynamically adapt TM schemes. By detecting stress points and high-traffic areas, the systems prevent infrastructure components from being overloaded, minimizing wear and tear and extending their lifespan [54].

The AI-powered pavement monitoring evaluates road surface conditions by analyzing data from thermal cameras, drones, and vibration sensors. The AI-reliant systems forecast pothole development, surface cracking, and rutting, enabling timely repairs that prevent additional damage [55]. They detect road wear and tear to plan preventive maintenance and evaluation of pavement condition.

5.4 AI-based Fault Detection in Critical Infrastructure

The AI-based systems continually monitor the structural health of bridges and tunnels using data from accelerometers, strain gauges, and corrosion sensors. The systems identify micro-cracks, rust, and deformations caused by loads that can lead to structural collapse [56]. The features are detection of bridge cracks and vibrations, prediction of tunnel wear and subsidence, and automated alert mechanisms.

The AISs scan rail track sensor information to identify wear and tear, misalignment, and derailment hazards. These systems predict track deterioration, enabling railway officials to take pre-emptive actions for passenger safety [57].

5.5 Maintenance Scheduling and Resource Optimization

Maintenance scheduling is automated through AISs that forecast infrastructure degradation and calculate optimal intervals for maintenance. The systems examine past maintenance history, weather conditions, and traffic volume data to create a ranked maintenance schedule [58].

Decision models driven by AI optimize the utilization of resources, including workforce, equipment, and materials, for effective maintenance activities. Through precise forecasting of maintenance requirements, AI models lower unnecessary expenses and improve operational effectiveness [59].

6. AI in Emergency Response and Post-Accident Analysis

6.1 Introduction

Integration of AI has led to a faster and more efficient emergency response system, enabling real-time decision-making [60].

The AI-based solutions complement emergency response, providing automated identification of accidents, a rapid deployment of resources, and smart routing of emergency vehicles. In addition to this, post-accident analysis through AI, such as by using video analytics and ML, allows the authorities to examine the causes of the accidents, improve safety measures, and avoid a second incident [61]. By analyzing information recorded by surveillance cameras, vehicle sensors, and innovative infrastructure, AI-based systems can provide a detailed account of accidents and help predict collision patterns, identifying areas of risk [62]. AI solutions offer a better level of preparation, and they remove the chances of human failure during emergencies [63].

6.2 AI in Emergency Response Systems

The AI-dispatch systems process incident evidence and dynamically deploy ambulances, fire crews, and police to incident sites. AI algorithms optimize routes depending on traffic and projected arrival time to ensure the fastest response [64].

The AI techniques utilized include:

- *Route Optimization Algorithms*: They optimize travel time and reduce congestion.
- *Predictive Resource Allocation*: It calculates resources needed according to accident severity.
- *Dynamic Traffic Signal Control*: It ensures seamless passage for emergency responders.

Drone AI algorithms provide emergency personnel with overhead photos and live video streams to assess accidents instantly, analyzing these photos to identify hazards, quantify damage, and aid decision-making [65].

6.3 AI for Severity Prediction and Triage Optimization

Deep learning and random forest classifiers process real-time information to forecast accident severity. Based on the speed of a vehicle, the angle of collision, and the type of collision, these models help determine the magnitude of injury and the need for medical treatment [66].

Accident victims are assessed through a triage system and based on the severity of their injuries and the availability of resources, the right treatment protocols are suggested so that the severely injured patients get quick attention [67].

The techniques employed include:

- *Decision Trees and Random Forests*: They help assign patients to categories based on injury severity.
- *Bayesian Networks*: They predict the likelihood of fatal injuries.
- *RL Models*: They make optimal triage decisions in a changing environment.

6.4 Post-Accident Analysis and Forensics

Artificial intelligence–driven video and sensor data are needed to recreate collisions in the future. AI models utilize traffic footage, dashcam recordings, and black box data to mimic accident setups and analyze their causes [68].

The AI algorithms link accident data to environmental, vehicular, and human elements to analyze root causes. AI models identify high-risk areas and suggest accident prevention [69].

6.5 AI for Behavioral Analysis and Driver Profiling

The AI models examine driver behavior based on information gathered from in-car telematics, GPS, and mobile phone apps. The models identify dangerous driving patterns, including abrupt braking, hard acceleration, and drowsy and fatigued driving, which can lead to accidents [70].

The AI-powered profiling systems determine high-risk drivers and vehicles based on past data and behavioral patterns. They provide this information to help insurance providers, fleet owners, and police authorities take early action in reducing risks [71].

7. Conclusion

The incorporation of AI into traffic safety systems changes transportation management and protection. AI technologies offer real-time decision-making, predictive modeling, and situational awareness, improving traffic control, accident prediction, infrastructure construction, and emergency response. ML, CV, and deep neural networks have transformed the TM paradigm, making roadways safer and more efficient using AI. Leading AI traffic safety apps include autonomous monitoring and management. CV and sensor networks help AISs detect anomalous traffic behavior and predict traffic conditions, dynamically altering signal timings. These technologies expedite traffic flow, notify authorities, and re-route vehicles in emergencies. AI can also conceptualize and forecast potential mishaps based on historical data and sensor readings, thereby helping to prevent collisions.

Other than improving real-time operations, AI is crucial to predictive maintenance. Data from infrastructure and vehicle sensors enable AI models to identify maintenance needs and to warn of potential safety issues. This reduces downtime, maintenance, and road hazards. AI-powered autonomous drones and analytics platforms are being utilized in emergency response due to their ability to detect accidents quickly, mobilize resources, and conduct forensic studies of accident causes.

Future technologies like quantum computing, federated learning, and DT implies Digital Twin simulations are expected to revolutionize traffic safety. These advancements could make AI applications more accurate, faster, and private. AISs, bias detection, bias reduction, and adversarial, resilient AI should be explained. AI in traffic safety has a bright future, but also faces obstacles. Success requires both technical skills and adherence to the above rules, ethical norms, and cooperative structures. AI's influence on social norms and regulatory frameworks can make AI systems safer, smarter, and more sustainable, establishing a futuristic transportation system.

References

[1] World Health Organization. (2019). *Global status report on road safety 2018*. World Health Organization.

[2] Fu, Y., Li, C. and Yu, F.R. (2021). A survey of driving safety with sensing, vehicular communications, and artificial intelligence-based collision avoidance. *IEEE Transactions on Intelligent Transportation Systems*, *23*(7), 6142–6163.

[3] Ebadinezhad, S., Stanley, K.A., et al. (2024). The impact of AI on traffic management and safety in the Internet of Vehicles through mobile edge computing. In *2024 International Conference on Inventive Computation Technologies*, 319–326.

[4] Estrela, V.V., Laghari, A.A., Lopes, R.T., Khan, A.A., Yin, S. and Deshpande, A. (2023). Introduction to intelligent healthcare in a post-pandemic world. In *Intelligent Healthcare Systems*, ed. V. V. Estrela, 3–23. CRC Press.

[5] Estrela, V.V. et al. (2023). Metaheuristics applied to pathology image analysis. In *Intelligent Healthcare Systems*, ed. V. V. Estrela, 322–340. CRC Press.

[6] Song, X., Zhang, Q., Sekimoto, Y. and Shibasaki, R. (2014). Intelligent system for urban emergency management during large-scale disaster. *Proceedings of the Twenty-Eighth AAAI Conference on Artificial Intelligence*, 458–464. https://ojs.aaai.org/index.php/AAAI/article/view/8758

[7] Agrahari, A., Dhabu, M.M. et al. (2024). Artificial intelligence-based adaptive traffic signal control system: A comprehensive review. *Electronics (Basel)*, *13*(19): 3875.

[8] Santos, D., Saias, J., Quaresma, P. and Nogueira, V.B. (2021). Machine learning approaches to traffic accident analysis and hotspot prediction. *Computers*, *10*(12), 157.

[9] Dr. Nirvikar Katiyar, Dr. Abhay Shukla, Dr. Namita Chawla, Mr. Raju Singh, Dr. Sudhir Kumar Singh and Mohd. Faraz Husain. (2024). AI in Autonomous Vehicles: Opportunities, Challenges, and Regulatory Implications. Educational Administration: Theory and Practice, 30(4), 6255–6264. https://doi.org/10.53555/kuey.v30i4.2373

[10] Estrela, V.V. et al. (2023). Remote sensing applications in disease mapping and public health analysis. In *Intelligent Healthcare Systems*, ed. V. V. Estrela, 185–202. CRC Press.

[11] Thrun, S. (2010). Toward robotic cars. *Commun ACM*, *53*(4), 99–106.

[12] Estrela, V.V., Hemanth, J. et al., eds. (2020). *Imaging and Sensing for Unmanned Aircraft Systems. Volume 1: Control and Performance*. The Institution of Engineering and Technology.

[13] Farfan, W.S., Saotome, O., Estrela, V.V. and Razmjooy, N.V. (2020). Integrated optical flow for situation awareness, detection and avoidance systems in UAV systems. In *Imaging and Sensing for Unmanned Aircraft Systems. Volume 1: Control and Performance*, eds V. V. Estrela, J. V. Hemanth, O. V. Saotome, G. V. Nikolakopoulos and R. V. Sabatini. The Institution of Engineering and Technologys.

[14] Dutta, V., Pawlicki, M., Kozik, R. and Choraś, M. (2022). Unsupervised network traffic anomaly detection with deep autoencoders. *Logic Journal of the IGPL, 30*(6), 912–925.
[15] Abdulhai, B., Pringle, R. and Karakoulas, G.J. (2003). Reinforcement learning for true adaptive traffic signal control. *Journal of Transportation Engineering, 129*(3), 278–285.
[16] Dahri, F.H. et al. (2024). Real-time detection of UAV small target's infrared imaging via OSTD-YOLOv8. *2024 IEEE URUCON*, 1–5.
[17] Zakaria, N.J., Shapiai, M.I., Abd Ghani, R., Yassin, M.N.M., Ibrahim, M.Z. andWahid, N. (2023). Lane detection in autonomous vehicles: A systematic review. *IEEE Access, 11*, 3729–3765.
[18] Deng, W. and Wu, R. (2019). Real-time driver-drowsiness detection system using facial features. *IEEE Access, 7*, 118727–118738.
[19] Tran, D.-K., Hoang, D.-Q. et al. (2022). Real-time traffic flow prediction using big data analytics. *International Conference on Intelligence of Things*, 398–405.
[20] Devlin, J., Chang, M.-W., Lee, K. and Toutanova, K. (2019). Bert: Pre-training of deep bidirectional transformers for language understanding. *Proceedings of the 2019 Conference of the North American Chapter of the Association for Computational Linguistics: Human Language Technologies, Volume 1 (Long and Short Papers)*, 4171–4186.
[21] Shukla, A.K., Kansal, P. and Parkash, J. (2024). Enhancing efficiency and functionality of voice-controlled cars through NLP techniques and additional features. *International Conference on Computing and Machine Learning*, 119–133.
[22] Mustapha, A., Saad, N., Mustapha, M., Raji Egigogo, A., Bala Umar, M., Jaya, S., Darul Ehsan, S.: Enhancing Road Safety Education through AI-Driven Tools: The Impact of Driving Simulations for Accident Prevention. (2025). (website: https://zenodo.org/records/15689692)
[23] Zhang, Y., Wang, Y. and Li, X. (2021). Reinforcement learning-based traffic signal control: A comprehensive review. *IEEE Transactions on Intelligent Transportation Systems, 22*, 3771–3787.
[24] Wen, Y., Lu, Y. and Yan, J. (2011). An algorithm for license plate recognition applied to intelligent transportation system. *IEEE Transactions on Intelligent Transportation Systems, 12*(3), 830–845.
[25] Ciuffo, B., Punzo, V. and Torrieri, V. (2008). Comparison of simulation-based and model-based calibrations of traffic-flow microsimulation models. *Transportation Research Record, 2088*(1), 36–44.
[26] Ponnusamy, S., Chourasia, H., Rathod, S.B. and Patil, D. (2024). "AI-driven traffic management systems: Reducing congestion and improving safety in smart cities. In *Smart Cities: Blockchain, AI, and Advanced Computing*, eds B. Sharma, M. Gupta and G. Jeon, 96–121. CRC Press.
[27] Jain, Y. and Pandey, K. (2025). Transforming urban mobility: A systematic review of AI-based traffic optimization techniques. *Archives of Computational Methods in Engineering*, 1–37.
[28] Djahel, S., Jabeur, N., Barrett, R. and Murphy, J. (2015). Toward V2I communication technology-based solution for reducing road traffic congestion in smart cities. *2015 International Symposium on Networks, Computers and Communications*, 1–6.
[29] Adewopo, V., Elsayed, N., Elsayed, Z., Ozer, M., Wangia-Anderson, V. and Abdelgawad, A. (2023). AI on the road: A comprehensive analysis of traffic accidents and accident detection system in smart cities. *arXiv preprint arXiv:2307.12128.*
[30] LeCun, Y., Bengio, Y. and Hinton, G. (2015). Deep learning. *Nature, 521*,(7553), 436–444.
[31] Deshpande, A.B., Estrela, V.V. and Khan, A. (2025). *Intelligent Designs, Innovations and Sustainability in Agriculture 4.0*. CRC Press.
[32] Deshpande, A., Estrela, V. v, Patavardhan, P., Kallimani, G.: Super-resolution in a world of scarce resources for medical imaging applications. In: Edge-AI in Healthcare. pp. 183–196. CRC Press (2023).
[33] Deshpande, A., Estrela, V.V. and Razmjooy, N. (2021). *Computational Intelligence Methods for Super-resolution in Image Processing Applications*. Springer.

[34] Krizhevsky, A., Sutskever, I. and Hinton, G.E. (2012). Imagenet classification with deep convolutional neural networks. *Advances in Neural Information Processing Systems, 25.*
[35] Dominguez, R. and Onieva, E. (2011). LIDAR based perception solution for autonomous vehicles. *International Conference on Intelligent Systems Design and Applications*, 790–795.
[36] Albasrawi, R., Fadhil, F.F. and Ghazal, M.T. (2022). Driver drowsiness monitoring system based on facial landmark detection with convolutional neural network for prediction. *Bulletin of Electrical Engineering and Informatics*, *11*(5), 2637–2644.
[37] Abou Elassad, Z.E., Mousannif, H., Al Moatassime, H. and Karkouch, A. (2020). The application of machine learning techniques for driving behavior analysis: A conceptual framework and a systematic literature review. *Engineering Applications of Artificial Intelligence*, *87*, 103312.
[38] Bojarski, M. et al. (2016). End to end learning for self-driving cars. *arXiv preprint arXiv:1604.07316.*
[39] Schwarting, W., Alonso-Mora, J. and Rus, D. (2018). Planning and decision-making for autonomous vehicles. *Annual Review of Control, Robotics, and Autonomous Systems*, *1*(1), 187–210.
[40] Outay, F., Kamoun, F., Kaisser, F., Alterri, D. and Yasar, A. (2019). V2V and V2I communications for traffic safety and CO2 emission reduction: A performance evaluation. *Procedia Computer Science*, *151*, 353–360.
[41] Tahir, M.N., Mäenpää, K. and Hippi, M. (2020). Pedestrian motion detection & pedestrian communication (P2I & V2P). *2020 International Conference on Software, Telecommunications and Computer Networks*, 1–3.
[42] Alahmed, Y., Abadla, R. and Al Ansari, M.J. (2024). Enhancing safety in autonomous vehicles through advanced AI-driven perception and decision-making systems. *Fifth International Conference on Intelligent Data Science Technologies and Applications*, 208–217.
[43] Paden, B., Čáp, M., Yong, S.Z., Yershov, D. and Frazzoli, E. (2016). A survey of motion planning and control techniques for self-driving urban vehicles. *IEEE Transactions on Intelligent Vehicles*, *1*(1), 33–55.
[44] Krizhevsky, A., Sutskever, I., Hinton, G.E.: ImageNet classification with deep convolutional neural networks. Commun ACM. 60, 84–90 (2017). https://doi.org/10.1145/3065386.
[45] Sarker, I.H. (2024). AI for critical infrastructure protection and resilience. In *AI-Driven Cybersecurity and Threat Intelligence: Cyber Automation, Intelligent Decision-making and Explainability*, 153–172. SpringerNature.
[46] Razmjooy, N., Deshpande, A., Khalilpour, M. and Estrela, V.V. (2020). Optimal bidding strategy for power market based on improved world cup optimization algorithm. In *Metaheuristics and Optimization in Computer and Electrical Engineering*, eds N. Razmjooy, M. Ashourian and Z. Foroozandeh, 137–152. Springer.
[47] Rane, N., Choudhary, S., Rane, J.: Artificial Intelligence (AI) and Internet of Things (IoT) - based sensors for monitoring and controlling in architecture, engineering, and construction: applications, challenges, and opportunities. SSRN Electronic Journal. (2023). https://doi.org/10.2139/ssrn.4642197.
[48] Estrela, V.V. et al. (2021). Motion estimation role in the context of 3D video. *International Journal of Multimedia Data Engineering and Management*, *12*(3), 16–38.
[49] Yuan, F.-G., Zargar, S.A., Chen, Q. and Wang, S. (2020). Machine learning for structural health monitoring: Challenges and opportunities. *Sensors and Smart Structures Technologies for Civil, Mechanical, and Aerospace Systems 2020*, *11379*, 1137903.
[50] Bono, F.M. et al. (2023). A deep learning approach to detect failures in bridges based on the coherence of signals. *Future Internet*, *15*(4), 119.
[51] Saikia, P., Sahu, B., Prasad, G. and Kumar, S. (2025). Smart infrastructure systems: A review of IoT-enabled monitoring and automation in civil and agricultural engineering. *Asian Journal of Research in Computer Science*, *18*(4), 24–44.

[52] Lam, D.D., Anh, L.D., Giang, L.T. and Ha, H. (2022). Risk identification and prediction for highway bridge projects using an artificial intelligence model. *International Conference on Sustainability in Civil Engineering*, 711–719.

[53] Estrela, V.V. et al. (2023). Digital twin framework for intelligent healthcare facilities through ISO/IEEE 11073. In *Intelligent Healthcare Systems*, ed. V. V. Estrela, 279–297. CRC Press.

[54] Rajendran, R.K., Blessing, N.R.W., Priya, T.M.: Traffic Flow Optimization Using AI. In: Recent Trends in Geospatial AI. pp. 217–238. IGI Global Scientific Publishing (2025). https://www.igi-global.com/book/recent-trends-geospatial/349908

[55] Feitosa, I., Santos, B. and Almeida, P.G. (2024). Pavement inspection in transport infrastructures using unmanned aerial vehicles. *Sustainability*, *16*(5), 2207.

[56] Macke, S., Munsch, S., Stascheit, J., Maidl, U. and Hegemann, F. (2024). AI-based anomaly detection in tunnel excavation data. *Geomechanics and Tunnelling*, *17*(4), 312–323.

[57] Dinesh, C., Kingson, A., Suresh, K.P. (2024). Smart rail track diagnostics and surveillance system for enhanced railway infrastructure maintenance and safety using IOT. *International Conference on Advanced Computing and Communication Systems*, 225–230.

[58] Maity, A., Zaremby, D., McMullen, D. and Gomez, J. (2011). Automated scheduling using condition based maintenance. *IEEE Conference on Prognostics and Health Management*, 1–8.

[59] Balasubramanian, S. (AI-driven solutions for sustainable infrastructure development and management. *International Journal of Artificial Intelligence in Engineering*, *2*(1), 1–11.

[60] Prasad, D.R.K.U., Subramanyam, K. and Malleswari, D.N. (2025). AI-driven emergency response system For vehicles: Enhancing safety and assistance. *Metallurgical and Materials Engineering*, May, 286–301.

[61] Basheer Ahmed, M.I. et al. (2023). A real-time computer vision based approach to detection and classification of traffic incidents. *Big Data and Cognitive Computing*, *7*(1), 22.

[62] Girdhar, M., You, Y. and Song, T.-J. (2022). Post-accident cyberattack event analysis for connected and automated vehicles. *IEEE Access*, *10*, 83176–83194.

[63] Trujillo, A., Orellana, M. and Acosta, M.I. (2019). Design of emergency call record support system applying natural language processing techniques. *Conference on Information Technologies and Communication of Ecuador*, 53–65.

[64] Dikshit, S., Atiq, A., Shahid, M., Dwivedi, V. and Thusu, A. (2023). The use of artificial intelligence to optimize the routing of vehicles and reduce traffic congestion in urban areas. *EAI Endorsed Transactions on Energy Web*, *10*, 1–13.

[65] Narayan, Y., Tarun, S., Ali, A. and Sharma, N. (2025). AI-enabled drone technology for disaster management: A review. *International Conference on Innovations in Bio-Inspired Computing and Applications*, 13–19.

[66] Tahernejad, A., Sahebi, A., Abadi, A.S.S. and Safari, M. (2024). Application of artificial intelligence in triage in emergencies and disasters: A systematic review. *BMC Public Health*, *24*(1), 3203.

[67] N. kuttan, A. Pundkar, C. Gadkari, A. Patel, and A. Kumar, "Transforming emergency medicine with artificial intelligence: From triage to clinical decision support", Multidiscip. Rev., vol. 8, no. 10, p. 2025285, Apr. 2025.

[68] Fang, J., Qiao, J., Xue, J. and Li, Z. (2023). Vision-based traffic accident detection and anticipation: A survey. *IEEE Transactions on Circuits and Systems for Video Technology*, *34*(4), 1983–1999.

[69] Chigurupati, A. and Lassar, N. (2017). Root cause analysis using artificial intelligence. *Annual Reliability and Maintainability Symposium*, 1–5.

[70] S. Yaqoob, G. Morabito, S. Cafiso, G. Pappalardo and A. Ullah, "AI-Driven Driver Behavior Assessment Through Vehicle and Health Monitoring for Safe Driving—A Survey," in IEEE Access, vol. 12, pp. 48044-48056, 2024, doi: 10.1109/ACCESS.2024.3383775.

[71] Kalla, D. (2022). AI-powered driver behavior analysis and accident prevention systems for advanced driver assistance. *International Journal of Scientific Research and Modern Technology*, pp: 1–9.

4 Artificial Intelligence for Measuring Transportation Comfort: A Comprehensive Review

Gabriel Caumo Vaz[1], Gabriel Gomes de Oliveira[2*],
Vitor Eduardo Molina Junior[3], William Machado Emiliano[4],
Yuri Alexandre Meyer[5], Anand Deshpande[6], Dhanashree Kulkarni[7] and
Vania V. Estrela[8]

Transportation comfort is a critical determinant of people's satisfaction while traveling (i.e., stakeholders who can be passengers, crew, or maintenance personnel) and their choice of transport mode. Yet its objective and comprehensive measurement remain significant hindrances. Traditional methods often rely on subjective surveys or limited objective metrics, failing to capture the multifaceted nature of human comfort perception in dynamic transportation environments. This chapter presents a detailed review of the application of artificial intelligence (AI) for measuring transportation comfort, addressing the limitations of conventional approaches. We explore how various AI techniques, including machine learning, computer vision, and sensor fusion, can integrate diverse data sources—ranging from physiological responses and environmental parameters to vehicle dynamics and passenger behavior—to provide a more holistic and accurate assessment of comfort. The chapter delves into the methodologies employed, the types of data utilized, and the potential for AI-driven systems to enhance urban planning, vehicle design, and overall passenger experience. This work emphasizes the transition from subjective assessments to data-driven, predictive models, highlighting the transformative potential of AI in creating more comfortable and sustainable transportation systems.

1. Introduction

Urbanization and global interconnectedness have imposed significant demands on transportation systems, engendering an ongoing quest for efficiency, safety,

1-2 School of Electrical and Computer Engineering, State University of Campinas, 400 Albert Einstein Ave., Cidade Universitária, Campinas, SP 13083-852, Brazil
3-5 School of Technology, University of Campinas (UNICAMP), 1888 Paschoal Marmo St., Jardim Nova Itália, Limeira, SP 13484-332, Brazil
6 Department of Artificial Intelligence and Data Science, Affiliated to Visvesvaraya Technological University (VTU), Angadi Institute of Technology and Management, Belagavi, India
7 Department of Computer Science & Engineering, Angadi Institute of Technology and Management, Belagavi, (Affiliated to Visvesvaraya Technological University (VTU)), Karnataka, India.
8 Telecommunications Department, Fluminense Federal University, Rio de Janeiro, Brazil.
* Corresponding author: ggomes@unicamp.br.

DOI: 10.1201/9781003648147-04

and comfort. Transportation comfort is often deemed as a subjective, intangible attribute and a crucial factor in passenger satisfaction, influencing mode selection, travel productivity, and overall well-being. A restful ride can turn a routine commute into a productive, relaxing experience, thereby significantly improving the perceived quality of life in urban settings. Conversely, discomfort may lead to stress, fatigue, decreased productivity, and potential health complications among frequent users [1].

The complex nature of human comfort perception in dynamic transportation environments presents a considerable challenge for its objective and comprehensive measurement. Individuals (stakeholders) utilizing transportation may include passengers, crew members, or maintenance staff. This chapter designates them as stakeholders. Comfort encompasses more than the absence of physical discomfort; an intricate interaction among physical, biological, and psychological factors forges it. Factors include exposure to vibrations, noise, lighting settings, thermal environment (temperature and humidity), air quality, seating ergonomics and spatial availability (crowding), and the psychological impact of perceived safety, reliability, and information provision [2, 3, 4, 5].

Traditionally, the assessment of transportation comfort has relied primarily on two approaches: direct subjective feedback and limited objective physical measurements. Subjective methods, such as questionnaires, interviews, and passenger diaries, aim to capture individual perceptions and preferences [6, 7]. While these methods provide valuable qualitative insights into passenger experiences, they inherently suffer from several limitations. These include subjectivity bias, reliance on memory recall (which can be inaccurate or incomplete), social desirability bias (where respondents may provide answers they believe are expected), and the inability to capture real-time, instantaneous fluctuations in comfort levels [8]. Moreover, interpreting and aggregating qualitative data can be complex and time-consuming.

On the other hand, objective physical measurements have typically focused on quantifiable parameters like vibration levels (often measured by accelerometers according to standards like ISO 2631-1), noise levels (decibels), and basic thermal conditions [9]. While these provide foundational data, they often fail to encapsulate the holistic experience of comfort. For instance, a vehicle might exhibit low vibration levels but still be perceived as uncomfortable due to excessive noise, poor air quality, or a sense of overcrowding [10]. Furthermore, isolating these parameters does not account for their complex interactions or their cumulative impact on human perception. The disconnect between these isolated objective metrics and the holistic subjective experience highlights a critical gap in traditional comfort assessment.

The arrival of artificial intelligence (AI) and its rapid advancements offer a transformative paradigm for addressing these limitations in measuring transportation comfort [11, 12, 13, 14]. AI, encompassing powerful techniques such as machine learning (ML), deep learning, computer vision, and sensor fusion, provides unparalleled capabilities for processing vast amounts of heterogeneous data, identifying complex, nonlinear patterns, and making accurate predictions [11, 13, 15]. By leveraging AI, researchers and engineers can move beyond static, post-hoc, and fragmented assessments to develop dynamic, real-time comfort monitoring and prediction systems. This enables the integration of diverse data sources—including vehicle performance data, environmental parameters, physiological responses from wearable sensors, and even visual data of passenger behavior—to construct a more comprehensive and accurate picture of passenger comfort [11, 16, 17].

This chapter aims to provide a comprehensive review of the application of AI in measuring transportation comfort. It will delve into the methodological approaches employed, the diverse types of data utilized, the specific AI techniques that are most relevant, and the significant potential for AI-driven systems to revolutionize urban planning, vehicle design, operational management, and ultimately, the overall passenger experience. By highlighting the transition from subjective and fragmented assessments to data-driven, predictive models, this work underscores the transformative impact of AI in creating more comfortable, efficient, and sustainable transportation systems for the future.

The chapter unfolds as follows: the second section presents some methodology for transport comfort. The third section discusses applied AI to transport comfort. Future work appears in the fourth section. Finally, the fifth section concludes.

2. Methodology

The concept of transportation comfort, while seemingly intuitive, has evolved significantly as a field of scientific inquiry, shifting from rudimentary physical assessments to complex multi-factorial analyses. Early research in transportation comfort largely emerged from the fields of ergonomics, human factors, and mechanical engineering, primarily focusing on objective physical parameters [18]. The initial emphasis was on quantifying measurable aspects of the vehicle environment that directly impact human sensation.

A cornerstone of these early efforts was the study of vibration and shock. Researchers recognized that excessive vibrations could lead to passenger discomfort, fatigue, and even motion sickness. This led to the development of standardized measurement techniques and thresholds. For instance, ISO 2631-1, titled "Mechanical vibration and shock—Evaluation of human exposure to whole-body vibration—Part 1: General requirements," became a fundamental reference for assessing the effects of vibration on human health and comfort [9, 19]. Studies meticulously analyzed vibration frequency and amplitude across different levels of discomfort, often using accelerometers to quantify vehicle oscillations [9].

Beyond vibration, other environmental factors were identified as significant contributors to comfort. Research into noise levels within vehicle cabins became crucial, leading to efforts in acoustic design and sound insulation. Similarly, thermal comfort, encompassing temperature and humidity, was extensively studied, drawing upon principles from Heating, Ventilation, and Air Conditioning

(HVAC) engineering and human physiology. The Predicted Mean Vote (PMV) and Predicted Percentage of Dissatisfied (PPD) indices, developed by Fanger for indoor environments, have been widely adopted in transportation applications to quantify thermal sensation and discomfort. Research also began to consider air quality, acknowledging the impact of pollutants and ventilation on passenger well-being [17].

However, as research progressed, it became increasingly evident that a purely objective, physical approach to comfort measurement was insufficient. Quantifiable physical parameters do not solely dictate passengers' perceptions of comfort but are deeply influenced by subjective and psychological factors [2, 3, 20]. This realization marked a pivotal shift in the bibliographic landscape. Studies began to incorporate qualitative methods, such as questionnaires, interviews, and focus groups, to gather passengers' direct feedback on their experiences. This revealed that factors like crowding levels, the quality of interior aesthetics, the sense of personal space, perceived safety and security, and even the provision of information (e.g., real-time updates on delays) profoundly influenced overall comfort and satisfaction [2, 3, 4, 20]. The discrepancies often observed between objective measurements (e.g., low vibration levels) and subjective perceptions (e.g., high reported discomfort) underscored the need for a more holistic understanding [6, 7, 20]. However, these subjective methods, while rich in qualitative data, posed challenges in scalability, real-time application, and susceptibility to biases such as recall inaccuracy and social desirability.

The recent confluence of advancements in sensor technology, the explosion of big data, and the unprecedented capabilities of AI has ushered in a new era for comfort measurement. AI provides the computational power and analytical frameworks necessary to bridge the long-standing gap between objective physical parameters and subjective human perception [10, 11]. The main areas where AI can improve transport comfort are discussed next.

2.1 Personalized In-Vehicle Experience

By using conversational voice assistants for hands-free operation, remembering driver preferences to adjust settings (e.g., climate and seat position), and providing tailored recommendations for infotainment, dining, and navigation, AI personalizes the in-car experience. To improve efficiency, safety, and convenience, it also offers cutting-edge technologies like driver behavior analysis, real-time traffic information, and predictive maintenance.

2.2 Innovative Traffic and Route Management

Innovative Traffic and Route Management (ITRM) systems rely on technologies such as AI, sensors, and connected vehicles for surveillance and control. ITRMs monitor traffic in real time to integrate operational data (e.g., vehicle speed profiles, braking patterns, and route characteristics), optimize signal timing, and provide dynamic route suggestions. Key innovations include AI-powered adaptive traffic signals and predictive analytics to anticipate congestion in connected vehicle systems that communicate with each other and infrastructure, all aimed at optimizing route planning. These solutions improve efficiency, reduce delays and emissions, and enhance safety [21, 22].

2.3 Extensive Multimodal, Multisensor Fusion

The application of AI in transportation comfort influences its ability to process vast, heterogeneous datasets stemming from a plethora of sensing modalities [13, 14, 15]. Modern research now integrates data from an array of sophisticated sensors.

Accelerometers and gyroscopes not only measure vibrations but, when processed by AI, can also identify complex indicators of motion sickness or predict rough-ride segments [15, 23].

Environmental sensors (temperature, humidity, CO2, particulate matter, noise) feed into AI models that can infer thermal discomfort or air quality concerns more accurately than simple thresholds [16, 17].

The proliferation of on-body and in-body (e.g., ingestibles, implantables) wearable physiological sensors (e.g., smartwatches, chest straps) [24, 25] has enabled the collection of data such as heart rate variability (HRV), skin conductance, and even electroencephalography (EEG). AI algorithms can analyze these physiological responses as objective proxies for stress, relaxation, or discomfort, providing a direct link to the human body's response [26].

Computer vision techniques, powered by deep learning, can analyze passenger density to infer crowding levels, monitor posture, and even detect facial expressions (with strict ethical and privacy guidelines) to gauge perceived comfort or distress [13, 27]. ML models, including neural networks, support vector machines, random forests, and gradient boosting machines, are particularly adept at identifying complex, nonlinear relationships between these diverse inputs and perceived comfort levels. For instance, deep learning models can learn intricate patterns from raw time-series data (e.g., continuous vibration and noise) to more accurately predict passenger comfort than traditional statistical methods [28]. The ability of AI to perform sensor fusion—combining data from multiple disparate sources [29, 30, 31, 32]—is key to creating a truly holistic comfort assessment system [33]. This advancement moves the field from fragmented insights to integrated, real-time, and predictive comfort analytics, thereby addressing the limitations of both purely objective and purely subjective historical approaches.

3. AI as a Tool for Coordinating Terrestrial, Water, Aerial, and Spatial Vehicles

3.1 Ergonomic Impacts of Multiple Vehicles Working Simultaneously

The ergonomic effects of having cars bicycles, and motorcycles in a shared environment (such as public roads, excluding the specific environment of an airfield for aircraft operations) primarily stem from the significant variations in vehicle size, speed, maneuverability, and protection offered to users, leading to heightened risks and increased cognitive load for all involved [34–39]. The key consequences include:

A. Bigger safety and physical risks due to the vulnerability of road users. By vulnerable road users (VRUs) one means cyclists and motorcyclists lacking the physical protection of a car, making them significantly more susceptible to severe injury or fatality in a collision.
B. Increased collision risk stemming from differences in vehicle dynamics, braking capabilities, and visibility can lead to accidents. Stakeholders may accept smaller time gaps when approaching motorcycles than when approaching cars, a risky behavior that increases the danger of a crash.
C. Augmented physical injury potential for motorcyclists, continuous riding, especially with poor posture due to design constraints, can lead to musculoskeletal disorders (MSDs), back pain, and discomfort in the arms, elbows, wrists, and thighs.
D. High environmental exposure, since motorcyclists and bicyclists are directly exposed to harsh weather conditions (cold, rain) and higher levels of air pollution from nearby motor vehicles, can affect comfort and health, though the health benefits of physical activity for cyclists generally outweigh the pollution risks.
E. High cognitive and perceptual ergonomics due to increased cognitive load, as participants across all vehicle types, particularly in mixed traffic, constantly monitor a wide range of vehicle sizes, speeds, and behaviors. This includes judging the distance and arrival time of various vehicles, a complex task that can be prone to human error.
F. Manifold visibility issues because VRUs are smaller and harder to see in traffic, leading other drivers to miss them, placing an ergonomic burden on VRUs to make themselves more visible (e.g., using signaling lights, wearing bright-colored apparel) and on drivers to be more vigilant.
G. More communication challenges arise from the need to establish effective communication among road users, which is critical for safety, but can be difficult in a heterogeneous traffic environment due to differences in vehicle design and user expectations.

Summarizing, the primary ergonomic consequence is the creation of a high-risk, high-cognitive-load environment due to the wide disparity among vehicle types. This highlights the need for dedicated infrastructure (e.g., physically separated bicycle lanes) and improved regulation to manage these interactions safely.

3.2 Manifold Vehicle-type AI Traffic Management

Artificial intelligence is crucial for managing diverse autonomous vehicles in shared urban and extra-urban spaces by enabling real-time decision-making, navigation, object recognition, and coordination among different modes of transport.

Autonomous cars, trucks, and delivery robots may use AI for navigation and to share roads with human-driven vehicles, pedestrians, and cyclists while connected and autonomous vehicles (CAVs) use vehicle-to-vehicle (V2V) and vehicle-to-infrastructure (V2I) communication to share real-time traffic data, optimize routes, and reduce congestion.

Uncrewed aerial vehicles (UAVs), or drones, are increasingly used in urban environments for package delivery, surveillance, environmental monitoring, and emergency response. AI enables drones to fly along planned routes without human intervention, using computer vision and deep learning to navigate complex urban areas and avoid obstacles like buildings and other aircraft. AI-coordinated swarm intelligence allows multiple drones to work as a team, sharing information and movements to accomplish tasks more efficiently than a single unit. Real-time decision-making at the edge computing layer enables on-board AI to process data and make independent decisions, which is crucial in areas with limited network connectivity.

Autonomous maritime or fluvial vehicles (boats) are liquid-surface vessels being developed and tested for use in urban waterways (e.g., canals and seashores) for transportation and infrastructure monitoring. Obstacle avoidance can rely on AI-powered perception systems that use Euclidean clustering and multi-object tracking to detect and track static and dynamic obstacles (e.g., other boats, debris) in the water. Efficient path planners, often based on nonlinear model predictive control (NMPC) schemes, calculate optimal, collision-free trajectories.

Extra-urban/global vehicles (spacecraft and satellites), while not directly "urban," share an extra-urban space (Earth's orbit and beyond) and rely on similar AI principles. Autonomous operations employing AI allow satellites to make independent decisions, such as adjusting orbits to avoid collisions with space debris, without constant human oversight. Mission management may use AI to automate and optimize complex space missions, handling large volumes of telemetry data and enabling space vehicles to operate efficiently and adapt to environmental conditions.

Overall coordination of different types of vehicles across different environments requires tools used in self-driving car development, which are being applied in satellite mission control, highlighting commonalities in managing high volumes of sensor data and autonomous operations.

The future vision involves interconnected, multimodal transportation networks where a central AI system manages the seamless flow of vehicles across domains (ground, air, water, space). This requires robust data sharing, standardized communication protocols, and complex AI algorithms to ensure safety, efficiency, and sustainability.

The application of AI in measuring transportation comfort represents a significant leap forward, transforming a largely subjective and qualitative assessment into a more objective and data-driven science. The ability to integrate multimodal sensor data and leverage advanced analytical techniques allows for a more nuanced understanding of what constitutes "comfort" in dynamic transportation environments.

One of the primary advantages is the shift from reactive to proactive comfort management. Instead of relying on passenger complaints after a journey, AI systems can detect discomfort in real time and potentially mitigate it. This has profound implications for improving passenger satisfaction, encouraging public transportation use, and enhancing the overall quality of life in urban areas.

However, several challenges and ethical considerations need to be addressed. Data privacy, especially when collecting physiological or visual data from passengers, is paramount. Robust ethical guidelines and transparent data usage policies are essential. The interpretability of complex AI models can also be a challenge; understanding why an AI predicts a certain comfort level is crucial for actionable insights and continuous improvement. Furthermore, the generalizability of AI models trained on specific vehicle types or routes to diverse transportation scenarios remains to be researched.

4. Comfort Metrics

Some metrics that enable progress in this area are discussed in this section.

4.1 Feature Engineering

Raw sensor data, while rich, often needs to be pre-processed and transformed into meaningful features that AI models can effectively learn from. This step is critical for improving model performance and interpretability.

- *Time-Domain Features*: For dynamic data (vibration, noise), statistical measures over defined time windows (e.g., 1-second or 5-second intervals) are commonly extracted:
    - Mean, Median, Standard Deviation: Represent average levels and variability.
    - Root Mean Square (RMS): A common metric for vibration magnitude [9].
    - Peak-to-Peak Amplitude: Indicates maximum excursion of the signal.
    - Kurtosis and Skewness: Describe the shape of the data distribution, highlighting impulsiveness or asymmetry, particularly useful for characterizing shocks.
    - Jerk: The rate of change of acceleration, a critical parameter for ride quality, derived from acceleration data.
- *Frequency-Domain Features*: Applying signal processing techniques like the Fast Fourier Transform (FFT) to time-series data (e.g., vibration, noise) allows for analysis in the frequency domain.
    - Power Spectral Density (PSD): Reveals the distribution of signal power across different frequencies, identifying dominant frequencies that are particularly annoying or comfortable [19].
    - Frequency Band Powers: Summing power within specific frequency bands (e.g., human-perceivable vibration ranges: 0.5–80 Hz; speech frequency ranges for noise) [9].

- *Derived and Composite Metrics*:
  - Comfort Indices: Calculation of established indices like Effective Vibration Dose Value (EVDV), Motion Sickness Incidence (MSI), or adapting PMV/PPD for transportation environments [19, 40].
  - Interaction Features: Creating new features that represent the interaction between different sensors (e.g., noise level and vibration amplitude), as combined stimuli often influence human perception.
  - Contextual Features: Incorporating non-sensor data like time of day, day of week, route segment, traffic density, and weather conditions, as these can indirectly influence perceived comfort.

4.2 AI Model Selection and Training

The choice of AI model depends heavily on the problem type (regression for continuous comfort scores, classification for comfort levels), the nature of the data, and computational resources.

- *Supervised Learning for Comfort Prediction*: The most common approach, where the model learns a mapping from input features to known comfort labels (either continuous scores or discrete categories).
- *Regression Models*: For predicting a continuous comfort score (e.g., 0–10 scale).
  Algorithms include:
  - Linear Regression: Simple, baseline model.
  - Support Vector Regression (SVR): Effective for high-dimensional data, less sensitive to outliers.
  - Ensemble Methods: Random Forests, Gradient Boosting Machines (GBM) (e.g., XGBoost, LightGBM) are powerful, robust, and handle complex nonlinear relationships and interactions between features very well [17, 19]. They often provide high accuracy by combining predictions from multiple decision trees.
- *Classification Models*: For categorizing comfort into discrete levels (e.g., "Very Comfortable," "Comfortable," "Slightly Uncomfortable," "Uncomfortable").
  Algorithms include:
  - i. Support Vector Machines (SVMs): Effective for clear separation boundaries.
  - ii. Decision Trees/Random Forests: Provide interpretable rules.
  - iii. K-Nearest Neighbors (KNN): Simple, instance-based learning.
  - iv. Neural Networks (NNs)/Multilayer Perceptrons (MLPs): Capable of learning complex nonlinear patterns.
- *Deep Learning for Raw Data Analysis and Complex Patterns*:
  - Convolutional Neural Networks (CNNs): Particularly effective for image-based data (e.g., from computer vision for crowd detection or posture analysis) and can also be applied to time-series data (1D CNNs) for automatically extracting relevant features without explicit feature engineering [28].

- Recurrent Neural Networks (RNNs)/Long Short-Term Memory (LSTM) Networks: Ideal for sequential time-series data (e.g., continuous streams of vibration, noise, or physiological data) as they can learn temporal dependencies and long-range correlations, which are crucial for understanding dynamic comfort changes [28].
- Transformers: Emerging deep learning architectures, initially developed for natural language processing but increasingly applied to time-series data, are showing promise for efficiently capturing long-range dependencies.

- *Unsupervised Learning for Pattern Discovery*:
  - Clustering Algorithms (e.g., K-Means, DBSCAN): Can identify natural groupings or segments of journeys with similar comfort characteristics in unlabeled data, which might reveal previously unknown comfort states or passenger clusters with shared sensitivities.

Dimensionality Reduction Techniques (e.g., Principal Component Analysis (PCA), t-SNE): Used to reduce the number of features while retaining most of the variance, simplifying models and identifying the most influential comfort factors.

Reinforcement Learning (RL) for Adaptive Systems: For systems that aim to dynamically optimize comfort. RL algorithms can learn optimal control policies (e.g., for active suspension, smart HVAC systems) by trial and error, based on rewards (e.g., high comfort scores) received from the environment [22].

- *Training Process*: The chosen model is trained on a substantial, carefully curated dataset where input features are mapped to known comfort labels (for supervised learning). This involves:
  - Hyperparameter Tuning: Optimizing model parameters (e.g., learning rate, number of layers/trees) using the validation set.
  - Loss Function Minimization: The model learns by minimizing a loss function (e.g., Mean Squared Error for regression, Cross-Entropy for classification) during an iterative optimization process (e.g., gradient descent).
  - Cross-Validation: K-fold cross-validation techniques are often used to ensure the model's robustness and generalization ability across different subsets of the data, reducing overfitting.

4.3 Model Evaluation and Validation

After training, the AI model's performance must be rigorously evaluated to ensure its accuracy, reliability, and practical applicability [41].

- *Performance Metrics*:

  For Regression Models (continuous comfort score):
  - Mean Absolute Error (MAE): Average absolute difference between predicted and actual values.
  - Root Mean Squared Error (RMSE): Measures the average magnitude of errors, penalizing larger errors more.

    - R-squared (R^2): Indicates the proportion of variance in the dependent variable that is predictable from the independent variables. Higher R^2 suggests a better fit.
- *For Classification Models (Discrete Comfort Levels)*:
    - Precision, Recall, F1-Score: More nuanced metrics, especially for imbalanced datasets, reflecting the model's ability to avoid false positives and false negatives.
    - Confusion Matrix: A table showing the number of true positives, true negatives, false positives, and false negatives.
    - ROC (Receiver Operating Characteristic) Curve and AUC (Area Under the Curve): Illustrate the classifier's performance across different threshold settings.
    - Validation against Subjective Assessments: A crucial step involves comparing the AI model's predictions with independent subjective comfort ratings collected from passengers. This helps confirm whether the objective, AI-driven metrics genuinely align with human perception of comfort. Statistical correlation analyses (e.g., Pearson's correlation coefficient) between predicted and reported comfort scores are commonly used [10].
- *Robustness and Generalizability Testing*:
    - Testing on Unseen Data: The model's performance on the dedicated test set (data not used in training or validation) is the primary indicator of its real-world effectiveness.
    - Cross-Modal Validation: Testing a model trained on one transportation mode (e.g., bus) on data from another (e.g., train) to assess generalizability, though often recalibration or fine-tuning is necessary.
    - Sensitivity Analysis: Examining how changes in individual input features affect the predicted comfort score, which can help in identifying the most influential comfort factors and guiding design improvements.

By meticulously following these methodological steps, researchers can develop robust, reliable AI systems capable of objectively measuring and predicting transportation comfort, thereby providing actionable insights to improve the passenger experience.

5. Future Works

The rapidly evolving landscape of AI and sensor technologies presents numerous exciting avenues for future research and development in transportation comfort measurement. These areas will contribute to refining current methodologies, expanding the scope of application, and addressing existing challenges:

- *Development of Standardized, XAI-based Comfort Metrics* [41, 42]: A critical next step is the establishment of widely accepted and standardized AI-driven comfort indices. This involves not only developing robust algorithms that correlate objective data with subjective comfort but also ensuring the explainability of these models via Explainable AI (XAI). Future research should focus on creating models that can articulate why a particular comfort level is predicted, providing actionable insights to engineers and urban planners. This will foster greater trust and facilitate the practical application of AI in industry.

- *Advanced Sensor Fusion and Novel Data Sources*: Future work should explore the integration of even more diverse and sophisticated sensor technologies. This includes miniaturized, non-invasive physiological sensors for continuous monitoring (e.g., smart textiles, embedded sensors in seats), advanced Lidar/radar for more precise environmental perception, and the use of edge computing for real-time data processing at the source. Research into leveraging new data sources like social media sentiment analysis (with privacy safeguards) or open-source weather data to contextualize comfort perceptions could also yield valuable insights.
- *Personalized and Adaptive Comfort Systems with RL*: Moving beyond general comfort predictions, future research should focus on developing truly personalized comfort profiles. This involves AI models that learn individual passenger preferences and sensitivities over time. Furthermore, the application of RL could enable self-optimizing transportation systems that continuously adjust vehicle parameters (e.g., suspension stiffness, HVAC settings, seat adjustability) in real time to maximize individual passenger comfort based on continuous feedback loops. This would transform static comfort into a dynamic, user-centric experience.
- *Integration with Autonomous Vehicle Design and Operation*: As autonomous vehicles become more prevalent, the role of AI in managing passenger comfort will become even more critical. Future work should investigate how AI can optimize ride quality, interior environment, and even the "sense of control" in a driverless context [43]. This includes researching novel vehicle interior designs that leverage AI to adapt to different passenger activities (work, relaxation, entertainment) and ensure seamless, comfortable transitions during autonomous maneuvers.
- *Longitudinal Studies and Long-term Impact Assessment*: Most current studies are relatively short-term. Future research needs to conduct extensive longitudinal studies to understand how passenger comfort perceptions evolve over longer periods and across repeated journeys. This will help in identifying chronic discomfort issues, assessing the long-term impact of comfort improvements on passenger well-being, and validating the sustained effectiveness of AI-driven interventions.
- *Ethical AI and Privacy-Preserving Techniques* [44]: Continued research is vital in developing robust ethical guidelines and implementing privacy-preserving AI techniques (e.g., federated learning, differential privacy) for comfort measurement systems. This will ensure that the benefits of AI are realized without compromising individual privacy or fostering surveillance concerns, thereby building public trust and facilitating widespread adoption.

- *Simulation and Digital Twin Technologies* [45]: The creation of high-fidelity digital twins of transportation environments and passenger physiology will enable cost-effective simulation and testing of AI comfort models. This allows for rapid prototyping and optimization of comfort-enhancing strategies without extensive physical deployments, accelerating innovation in the field.

By focusing on these areas, future research can further unlock the potential of AI to create truly comfortable, personalized, and sustainable transportation systems that cater to the diverse needs and preferences of all passengers [46, 47, 48].

6. Conclusion

The comprehensive exploration of AI's application in measuring transportation comfort unequivocally demonstrates a significant paradigm shift from traditional, often subjective, assessment methods to a more objective, real-time, and holistic approach. AI, through its powerful capabilities in data processing, pattern recognition, and predictive modeling, enables the integration of diverse data streams—ranging from vehicle dynamics and environmental parameters to physiological responses and passenger behavior. This multimodal data fusion, processed by sophisticated ML algorithms, offers an unprecedented understanding of the multifaceted nature of human comfort in dynamic transportation environments.

The key strength of AI-driven comfort measurement lies in its ability to move beyond reactive problem-solving to proactive intervention. By continuously monitoring and predicting comfort levels, AI systems can inform real-time adjustments to vehicle settings (e.g., suspension, climate control), optimize route planning, and even guide urban infrastructure development. This translates directly into enhanced passenger satisfaction, improved public transport ridership, and ultimately, the creation of more sustainable and human-centric urban mobility ecosystems. The capacity for personalization, where AI models can learn and cater to individual passenger preferences, further elevates the potential for truly bespoke and comfortable travel experiences. The economic benefits are also considerable, as increased comfort can lead to higher passenger loyalty, reduced operational costs through optimized resource allocation, and a stronger competitive edge for transportation providers.

While the transformative potential is immense, it is crucial to acknowledge and address the inherent challenges. Foremost among these are the ethical considerations surrounding data privacy, particularly when dealing with sensitive physiological and visual data of passengers. Robust frameworks for data collection, anonymization, and secure storage are paramount to building public trust. Furthermore, the "black box" nature of some complex AI models poses challenges in interpretability; understanding why an AI system identifies a particular level of discomfort is essential for engineers and designers to implement effective solutions. Finally, the generalizability of AI models across diverse transportation modes, operational conditions, and passenger demographics requires rigorous validation and continuous refinement to ensure their effectiveness in real-world scenarios.

In essence, the ongoing integration of AI into transportation comfort measurement is not merely an incremental improvement but also a fundamental redefinition of how we perceive, quantify, and ultimately enhance the passenger journey. It paves the way for a future where comfort is not an afterthought but an intrinsic, measurable, and optimizable component of every trip.

References

[1] Borck, R. (2019). Public transport and urban pollution. *Regional Science and Urban Economics*, *77*, 356–366.

[2] Friman, M., Lättman, K. and Olsson, L.E. (2020). Public transport quality, safety, and perceived accessibility. *Sustainability*, *12*(9), 3563.

[3] Imre, Ş. and Çelebi, D. (2017). Measuring comfort in public transport: A case study for Istanbul. *Transportation Research Procedia*, *25*, 2441–2449.

[4] Li, X.-H., Huang, L., Li, Q. and Liu, H.-C. (2020). Passenger satisfaction evaluation of public transportation using Pythagorean fuzzy multimoora method under large group environment. *Sustainability*, *12*(12), 4996.

[5] Shen, X., Feng, S., Li, Z. and Hu, B. (2016). Analysis of bus passenger comfort perception based on passenger load factor and in-vehicle time. *SpringerPlus*, *5*, 1–10.

[6] Esheti, S.A., Emagnu, Y.M., Haylemariam, B.D. and Melaku, R.S. (2024). Comparative analysis of public transportation comfort in Addis Ababa: Objective and subjective performance metrics. *Modelling and Simulation in Engineering*, *2024*(1), 2279130.

[7] Guastadisegni, G., De Pinto, S., Cancelli, D., Labianca, S., Gonzalez, A., Gruber, P. and Sorniotti, A. (2024). Ride analysis tools for passenger cars: Objective and subjective evaluation techniques and correlation processes—A review. *Vehicle System Dynamics*, *62*(7), 1876–1902.

[8] Shan, J., Bĕsinovi'c, N. and Schönberger, J. (2024).. *Journal of Rail Transport Planning & Management*, *29*, 100432.

[9] Zoccali, P., Loprencipe, G. and Lupascu, R.C. (2018). Acceleration measurements inside vehicles: Passengers' comfort mapping on railways. *Measurement*, *129*, 489–498.

[10] Griffin, M.J. (2007). Discomfort from feeling vehicle vibration. *Vehicle System Dynamics*, *45*(7–8), 679–698. https://doi.org/10.1080/00423110701422426

[11] Dhaliwal, A. (2022). Towards AI-driven transport and logistics. In *Digital Transformation in the Viral Age, WeB 2022, Lecture Notes in Business Information Processing*, eds A. Kathuria, P.P. Karhade, K. Zhao & D. Chaturvedi, 119–131. Springer, Cham.

[12] Kaledinova, E. and Boeldak, D. (2024). The assessment of the impact of artificial intelligence on transportation sustainability. *Zeszyty Naukowe. Transport/Politechnika Slkaska'*.

[13] Kolesnikova, I. (2024). *How AI in transportation can improve our everyday lives?* Mindtitan.

[14] Sadou, A.M. and Njoya, E.T. (2023). Applications of artificial intelligence in the air transport industry: A bibliometric and systematic literature review. *Journal of Aerospace Technology and Management*, *15*, e2223.

[15] Zantalis, F., Koulouras, G., Karabetsos, S. and Kandris, D. (2019). A review of machine learning and IoT in smart transportation. *Future Internet*, *11*(4), 94.

[16] Childers, B.A., Froggatt, M.E., Allison, S.G., Moore Sr, T.C., Hare, D.A., Batten, C.F. and Jegley, D.C. (2001). Use of 3000 Bragg grating strain sensors distributed on four 8-m optical fibers during static load tests of a composite structure. In *Smart Structures and Materials 2001: Industrial and Commercial Applications of Smart Structures Technologies*, Vol. 4332, 133–142.

[17] Franc, D. (2024). *Machine learning tools for shared vehicle air quality and comfort.*

[18] Griffin, M.J. (2012). *Handbook of Human Vibration.* Academic Press.

[19] Mansfield, N.J., Ashley, J. and Rimell, A.N. (2007). Changes in subjective ratings of impulsive steering wheel vibration due to changes in noise level: A cross-modal interaction. *International Journal of Vehicle Noise and Vibration*, *3*(2), 185–196.

[20] Cairns, S., Harmer, C., Hopkin, J. and Skippon, S. (2014). Sociological perspectives on travel and mobilities: A review. *Transportation Research Part A: Policy and Practice*, *63*, 107–117.

[21] Jevinger, ˚A., Zhao, C., Persson, J.A. and Davidsson, P. (2024). Artificial intelligence for improving public transport: A mapping study. *Public Transport*, *16*(1), 99–158.

[22] Lukic Vujadinovic, V., Damnjanovic, A., Cakic, A., Petkovic, D.R., Prelevic, M., Pantovic, V., ... and Bodolo, I. (2024). AI-driven approach for enhancing sustainability in urban public transportation. *Sustainability*, *16*(17), 7763.

[23] Kim, Y., Kwon, H., Kim, S., Kim, C. and Kim, T. (2003). Correlation of ride comfort evaluation methods for railway vehicles. *Proceedings of the Institution of Mechanical Engineers, Part F: Journal of Rail and Rapid Transit*, *217*(2), 73–88.

[24] Estrela, V.V., Grata, E.G.H., Deshpande, A., Sroufe, R., Lopes, R.T., Shi, F., Bukhsh, K., Rind, M. M. and Sarmad Shaikh, S. (2023). In-body devices and sensors communication—How implantables, ingestibles, and injectables interact with the Internet. In *Intelligent Healthcare Systems*, ed. V. V. Estrela, 236–258. CRC Press.

[25] Estrela, V.V., Intorne, A.C., Batista, K.K.S., Deshpande, A., Sroufe, R., Lopes, R.T., Shi, F., Yin, S. and Lin, Yu-Da (2023). Nanotechnology, Internet of nanothings and nanorobotics in healthcare—Nano for all. In *Intelligent Healthcare Systems*, V. V. Estrela, 259–278. CRC Press.

[26] Cacioppo, J.T., Tassinary, L.G., Berntson, G.G. et al. (2007). Psychophysiological science: Interdisciplinary approaches to classic questions about the mind. *Handbook of Psychophysiology*, *3*, 1–16.

[27] Pantic, M. and Rothkrantz, L.J.M. (2000). Automatic analysis of facial expressions: The state of the art. *IEEE Transactions on Pattern Analysis and Machine Intelligence*, *22*(12), 1424–1445.

[28] Du, X., Sun, C., Zheng, Y., Feng, X. and Li, N. (2021). Evaluation of vehicle vibration comfort using deep learning. *Measurement*, *173*, 108634.

[29] Tang, X., Zou, Y., Ma, J., Zhang, X., Liu, F. and Jiao, L. (2024). Multiple information collaborative fusion network for joint classification of hyperspectral and LiDAR data. *IEEE Transactions on Geoscience and Remote Sensing*, *62*, 1–16.

[30] Khan, A.A. et al. (2025). D-Agri: A solution for smart farming using AI & drone technology. *2025 IEEE 2nd International Conference on Electronics, Communications and Intelligent Science (ECIS)*, Yueyang, China, 1–6.

[31] Estrela, V.V., Deshpande, A., Lopes, R.T., Yin, S., Andreopoulos, N., Terziev, A. and Laghari, A. A. (2023). On DICOM, HEVC and 3D medical image compression for volumetric theragnostics. In *Intelligent Healthcare Systems*, ed. V. V. Estrela, 203–216. CRC Press.

[32] Lemonde, C., Arsenio, E. and Henriques, R. (2021). Integrative analysis of multimodal traffic data: Addressing open challenges using big data analytics in the city of Lisbon. *European Transport Research Review*, *13*(1), 64.

[33] Hall, D.L. and Jordan, J.M. (2010). *Human-centered Information Fusion*. Artech House.

[34] Doraswamy, B., Krishna, K.L. and Rao, T.C.S. (2024). A comprehensive review on real time traffic management coordination for smart cities with IoT connected drones. In *Proceedings of the 2024 International Conference on Cognitive Robotics and Intelligent Systems (ICC-ROBINS)*, Coimbatore, India, 17–19 April, 584–589.

[35] Mahboob, S. and Liu, L. (2023). Revolutionizing future connectivity: A contemporary survey on AI-empowered satellite-based non-terrestrial networks in 6G. *IEEE Communications Surveys & Tutorials*, *26*, 1279–1321.

[36] Zhang, X. (2025). Artificial intelligence in intelligent traffic signal control. *Applied and Computational Engineering*, *118*, 113–120.

[37] Moraga, Á., de Curtò, J., de Zarzà, I. and Calafate, C.T. (2025). AI-driven UAV and IoT traffic optimization: Large language models for congestion and emission reduction in smart cities. *Drones*, *9*(4), 248.

[38] Ara, J., Khan, M.B., Yasmin, S. and Bhuiyan, H. (2025). A review on the estimation of vulnerable road user behavior for automated vehicles. *International Journal of Sustainable Transportation*, *19*(6), 547–575.

[39] Baena, E., Testolina, P., Polese, M., Koutsonikolas, D., Jornet, J.M. and Melodia, T. (2025). Space-O-RAN: Enabling intelligent, open, and interoperable non-terrestrial networks in 6G. ArXiv, abs/2502.15936.

[40] Fanger, P.O. (1970). Thermal comfort. Analysis and applications in environmental engineering., (244 pp.), Copenhagen: Danish Technical Press.

[41] Estrela, V.V., de Jesus, M.A., Intorne, A.C., Batista, K.K.S., Deshpande, A., Shi, F., Laghari, A.A., Khan, A.A. and Oliveira, L.P. (2023) Medical visual theragnostic systems using artificial intelligence (AI)—Principles and perspectives. In *Intelligent Healthcare Systems*, ed. V. V. Estrela, 301–321. CRC Press.

[42] Romadlon, F., Saintika, Y., Aliyah, K. and Hamid, A. (2024). Projecting the needs of intelligent transportation system (ITS) based on comfort and security views of Trans Jateng's women ridership. *2024 International Conference on ICT for Smart Society (ICISS)*, 1–6.

[43] Dahri, F.H., Laghari, A.A., Dahri, N.A., Abro, G., Estrela, V.V., de Oliveira, G.G., Iano, Y. and Chuma, E.L. (2024). Real-time detection of UAV small target's infrared imaging via OSTD-YOLOv8. *2024 IEEE URUCON*, 1–5.

[44] Khan, A.A., Laghari, A.A., Almansour, H. et al. (2025). Quantum computing empowering blockchain technology with post quantum resistant cryptography for multimedia data privacy preservation in cloud-enabled public auditing platforms. *Journal of Cloud Computing*, *14*, 43.

[45] Estrela, V.V., Deshpande, A., Sroufe, R., Lopes, R.T., Grata, E.G.H., Andreopoulos, N., Terziev, A. and Laghari, A.A. (2023). Digital twin framework for intelligent healthcare facilities through ISO/IEEE 11073. In *Intelligent Healthcare Systems*, ed. V. V. Estrela, 279–297. CRC Press.

[46] Mohsen, B.M. and Mohsen, M. (2025). Reducing emissions through AI-driven multimodal transport optimization in IoT-connected environments. *International Journal of Energy Research*, (1), 2399288.

[47] Kale, A.A. (2025). Modern AI in green logistics: Smart systems for sustainability and carbon reduction. *Global Journal of Emerging AI and Computing*, *1*(2), 147–169.

[48] van Ardenne, M.T., Cebecauer, M., Cats, O. and Ma, Z. (2025). Personalised passenger information systems in public transport: A review and a 5-level personalisation taxonomy. *Transport Reviews*, *45*(6), 1016–1047.

5 AI-powered Driving Analytics: Understanding Driver Behavior through Data

Yash Shinde[1*], Prathmesh Redekar[2] and Prakash Sonwalkar[3]

Driver behavior analysis (DBA) is one of the most important research topics in the field of driving analytics. In this chapter, we review the current state of the art of artificial intelligence (AI)–driven driving analytics, which is a disruptive way to perceive driver behavior as the application of powerful technologies to analyze and process the tremendous amounts of information gathered by vehicle sensors, telematics, and connected devices. The AI-powered driving analytics is based on deep learning and machine learning frameworks, such as deep learning, neural networks, and deep learning algorithms, which are able to identify and define driver behaviors. The data collected from in-vehicle sensors, vehicle trajectories, and mobile data helps to create reliable driver behavior detection systems. In addition, machine learning–based analytics are implemented to help control fuel consumption, improve road safety, and offer immediate feedback to the driver, which results in a quantifiable increase in driving efficiency and reduction of accidents. Adaptive traffic management systems are AI-operated solutions to organize traffic flow and increase the efficiency of public transport, being dynamic in adjusting to the challenges of the cities.

1. Introduction

Artificial intelligence (AI)–powered driving analytics is an advanced way to understand driver behavior through the application of powerful technologies to analyze and process the tremendous amounts of information gathered by the vehicle sensors, telematics, and connected devices. These advanced systems utilize machine learning and AI solutions in tracking the real-time driving patterns, quantifying and assessing the driver safety and automobile effectiveness to ensure minimum costs of operation without compromising the security of employees and other road users. AI systems will be able to draft a detailed description of each driver as they continually monitor aspects of vehicle speed, acceleration, braking dynamics, and steering habits to compile a detailed profile of any particular driver

[1-3] Department of Computer Science and Engineering (AIML), Jain College of Engineering and Research, Affiliated to Visvesvaraya Technological University, Belagavi, Karnataka, India.

* Corresponding author: syash0080@gmail.com

DOI: 10.1201/9781003648147-05

and find areas that need improvement. The technology has allowed motorists instant feedback concerning their performance to correct risky behaviors before they result in accidents and has promoted the culture of safe and efficient driving.

1.1 Evolution of Vehicle Telematics and Connected Car Technologies

Digitization, wireless networks, and cloud services have been the key factors that have led to the evolution of vehicle telematics technologies and the connected car realm [1]. Vehicle-to-vehicle (V2V) and vehicle-to-infrastructure (V2I) communication is also implemented in connected vehicles, through which traffic safety and crash prevention are facilitated [2]. Dietmar P. F. Möller and Roland E. Haas combine cyber-physical systems (CPS) [3, 4] and the Internet of Things (IoT) to facilitate the transfer and exchange of information without any disturbances [5, 6, 7]. Within the confines of automotive telematics, there are multiple applications, such as GPS, infotainment, and onboard diagnostics, which are among the most effective to use for enhancing navigation and vehicle performance [8, 9]. New models and opportunities for business and services have emerged due to the development of connected cars, including car-sharing and smart ticketing [1]. With the advancement of this technology, self-driving cars are becoming a reality, which implies that they will affect many industries, such as the insurance field and transportation services [5, 8].

1.2 The Emergence of Big Data in Transportation Systems

The concept of big data analytics is transforming intelligent transportation systems (ITS) and connected/automated vehicles (CAV), providing innovative insights into traffic safety, driver behavior, and system maintenance [10]. Now, with the availability of information derived in different ways (structured and unstructured data produced by sensors, text analytics, and image processing), more holistic analysis and prediction of traffic patterns, accidents, and driver behavior can be undertaken [3, 11, 12]. Big data analytics also plays a critical role in the maintenance of the transportation infrastructure in terms of using the Internet of Things (IoT) and sensors to gather and analyze large volumes of data and make efficient decisions about its limitations and constraints [14].

1.3 Growing Importance of DBA for Safety and Efficiency

The increasing significance of driver behavior analysis (DBA) for both safety and efficiency, according to an ITS, is realized in most of the contemporary studies. DBA utilizes the information in in-vehicle networks, sensors, and communication technology to deliver driving patterns [15]. This analysis is applied to different things, such as traffic safety, the development of autonomous vehicles, energy management, and driver profiling [16]. Innovative platforms such as ADAPT DBA process data on board efficiently and add social incentives to enhance the safety of the drivers [17, 18].

1.4 Limitations

The limitations of subjective driver evaluation techniques are variability in human judgments [19, 20], lack of reproducibility [21], integration challenges with objective metrics, and need for balanced approach.

1.5 Need for Objective, Data-driven Insights into Driving Patterns

Hard data insights into driving trends are being considered more and more necessary in terms of road safety, improved fuel efficiency, and more civil driver behavior. Analytics with the help of AI offer a sound system for realizing these patterns by analyzing huge volumes of driving data. Machine learning AI technologies allow us to dynamically analyze the driving behavior, and as a result, make improvements that lead to fuel consumption and minimal environmental impact [22].

Data-driven frameworks and predictive models: In predicting driving behaviors, a data-driven framework based on unsupervised learning has been reported to be highly accurate as compared to the traditional approaches. In-vehicle sensor data could be used to split driving styles effectively and gather a pattern that can guide safety operations, particularly with elderly drivers [23, 24, 25].

1.6 Privacy and Ethical Considerations in Driver Monitoring

The ethical and privacy issues associated with AI-powered driving analytics and self-driving cars are known to be severe. The need to balance between safety and privacy is acute because the amount of data collected and relayed by the AI system about the driver is enormous [26, 27]. Through the adoption of responsible research and innovation frameworks, such problems could be mitigated by engaging the stakeholders and making driver monitoring socially acceptable [28, 29]. Algorithms of fairness and AI ethical use during development are vital in ensuring better decision-making and reduced risks to privacy.

1.7 Scope and Significance

Driving analytics applications use AI to analyze information from vehicle sensors, telematics, and highly sophisticated algorithms to evaluate and appreciate driver behavior. Some of the important uses of these analytics are:

- Insurance [30, 31, 32]
- Fleet management [33]
- Autonomous vehicle development [34–37]

In short, AI-based driving analytics are disrupting the automotive ecosystem with respect to safety, efficiency, and economic partnerships among insurers, fleet managers, technology providers, and their associates.

2. Methods

The specified data collection framework is aimed at recording a rich range of different driving behaviors, both with the help of innovative sensors and cheap data capture techniques, so that wide coverage is provided. Along with a solid data collection framework to ensure the appropriate analysis of driver behavior, AI-powered driving analytics is necessary. This framework incorporates numerous sensor technologies and sources of data that allow having detailed knowledge about driving patterns and behaviors [38], as shown in Figure 5.1. This design will take advantage of the smartphone sensors and other low-priced approaches to collect real-time data, improving the perception of driver behavior in naturalistic conditions [39, 40].

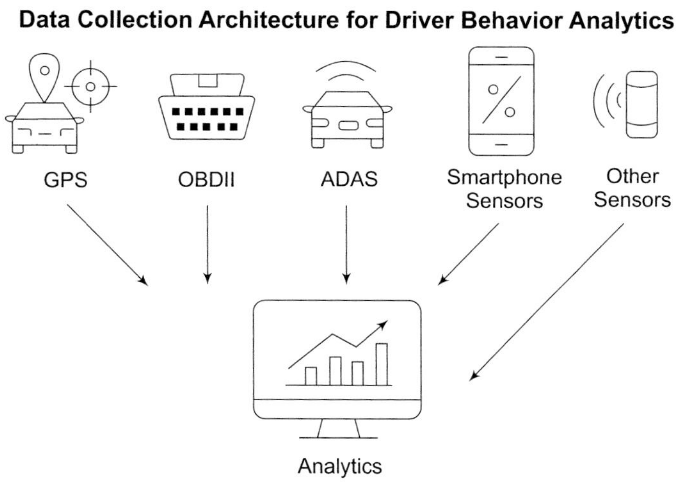

Fig. 5.1 Multimodal data collection architecture for driver behavior monitoring.

The frameowrk uses sensor technologies and data sources which are GPS Tracking Systems and Location Analytic [41], Measurements of Accelerometer and Gyroscope [42, 43], Data Guided Sensing Applications [44–46], Systems Integration, Advanced Driver Assistance [47], and Computer Vision (CV) [34, 49]. A data type is the mode of information delivery. It defines the extent to which information delivery can be conveyed. Data types are either binary, integer, or real. Binary data type is the codification of information. The combination of such a variety of data sources results in the diversity of data types and parameters necessary to conduct a complete driving habit analysis, including kinematic data, spatial data [50, 51], temporal patterns [52], and environmental context [53, 54, 55].

The starting point for effectively managing data quality is developing a sound framework that will help ensure that the data collected from various sources is accurate, complete, and timely. This is achieved by adopting strict data capture procedures and constant surveillance to detect any possible lapses in the quality process early enough [5].

- *Methods of Noise Filtration and Signal Filtering*: The smoothing of anomalies in the data presented by environmental conditions or inaccuracy in the operation of sensors is achieved by techniques like low-pass filters [57].
- *Imputation Strategies for Missing Data*: Missing data can highly distort the outcome of the analysis. To handle this, various methods can be used such as mean/mode imputation in a variable of numeric data, k-nearest neighbors, and regression-imputation strategies to impute missing information by using patterns [58].
- *The Procedures of Outlier Detection and Handling*: Outliers may be a representation of faulty data or rather one-of-a-kind habits of driving. The Z-score method and the interquartile range are some of the statistical tools that we apply to identify outliers [40, 60].

The key task of feature engineering is to transform raw data into usable variables that would improve the models. This involves behavioral pattern extraction algorithms [61], statistical feature computation [62], time-series analysis and segmentation [63], and spatial analytics and route optimization metrics [64].

2.1 AI and Machine Learning Approaches

It is critical to apply AI and machine learning approaches to develop meaningful interpretations of processed data and use them to support the right actions, as shown in Figure 5.2.

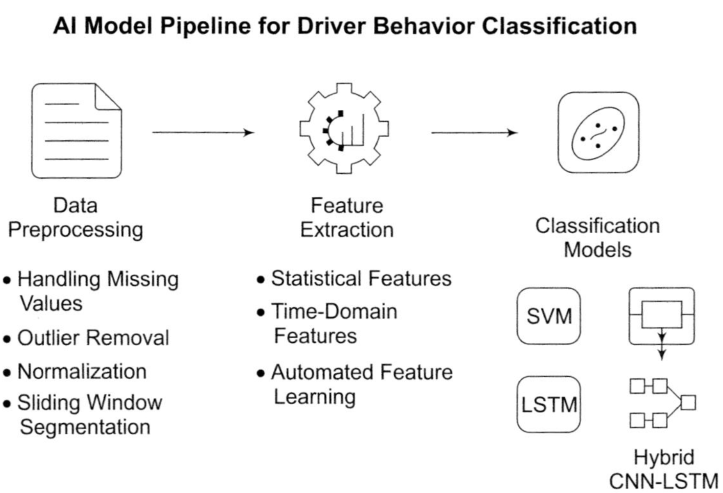

Fig. 5.2 Pipeline of AI-driven behavior analysis

2.2 Supervised Learning

The driver profiling models and risk prediction models are mainly based on supervised learning. The important methods include Classification Models to Profile Drivers [65], Regression on Risk Prediction [66], Support Vector Machine [67], and Spatial Analytics and Route Optimization Metrics [68].

2.3 Unsupervised Learning

The unsupervised learning methods are used to discover obscure data patterns without predetermined labels: Driver Segmentation with Clustering Algorithms [69] and Principal Component Analysis Dimensionality Reduction.

2.4 Deep Learning and High-Quality AI Techniques

Deep learning is used because of the potential to design very complex relationships within big datasets: Recurrent Neural Networks with Sequential Patterns, Long Short-Term Memory (LSTM) networks, transformer models, reinforcement learning, transfer learning, and explainable AI.

3. Analytics

3.1 Development of a Taxonomy of Behavior

A profound achievement of developing a detailed taxonomy of behaviors has been categorizing driver behavior into specific categories and achieving specific interventions. It has helped to detect cases of aggressive driving—such as quick acceleration, illegal braking, and frequent lane changes—through the use of strong machine learning algorithms.

3.2 Validating Performance and Verifications

The validations and the performance measures of these systems of classification can offer hopeful findings. The accuracies of classification vary across different drivers, but the precision and recall percentage rates of detecting behaviors are high in most cases.

3.3 Risk Assessment Models

The implementation of risk assessment models has yielded significant insights into accident probability prediction. By analyzing historical driving data and behavioral patterns, we have developed algorithms that predict the likelihood of accidents with a high degree of accuracy.

3.4 Behavioral Intervention Systems

Behavioral intervention systems have been designed to foster safer driving habits through real-time feedback mechanisms. These systems employ gamification strategies to encourage behavior modification, offering rewards for safe driving and personalized coaching recommendations tailored to individual drivers.

4. Discussion

The use of AI-powered driving analytics enables applying the data obtained from various sources to gain insight into the driving behavior that is concerned with the recognition of behavioral patterns. It implies determining important predictors of the driving behavior, the correlation analysis of different metrics, and the effect of the external factors. It also makes a difference between individual and population behavior analysis. The multimodal data analysis makes it possible to combine data collected from various sources and gain an overview of the driving behavior to identify complex relationships among metrics. Drive Sense enables the prediction of abnormal driving behaviors because it correlates findings with various driving patterns, thereby increasing the predictive capabilities.

5. Conclusion

The development of smart cities relies heavily on AI, with the connected infrastructure helping to provide data-based transportation services. Adaptive traffic management systems are AI-operated solutions to organize traffic flow, integrate smart city infrastructure, and increase the efficiency of public transport, being dynamic in adjusting to the challenges of the cities. The possibility of individual transportation optimization helps self-driving vehicles, focusing on the development of a collaborative ecosystem that fulfills individual mobility tasks. On the other hand, the fast development of AI in driving analytics can result in overdependence, reducing human driving capabilities and adversely affecting an individual's ability to be an adequate driver, thereby increasing their vulnerability in case of a system breakdown. It will be vital to strike a balance between innovation and caution to foster sustainability in this field. The operating field of AI-based driving analytics promises to achieve tremendous breakthroughs in the future, especially through fully automated driving systems, smart cities, customized commuting, and social implications.

References

[1] Arena, F., et al. (2021). Future connected cars through the evolution of telematics and infotainment. *International Journal of Engineering*, *9*, 49–61.

[2] Narla, S.R. (2013). The evolution of connected vehicle technology: From smart drivers to smart cars to self-driving cars. *ITE Journal: Institute of Transportation Engineers*, *83*, 2013.

[3] Estrela, V.V., de Jesus, M.A., Intorne, A.C., Batista, K.K.S., Deshpande, A., Shi, F., Laghari, A.A., Khan, A.A. and Oliveira, L.P. (2023). Medical visual theragnostic systems using artificial

intelligence (AI): Principles and perspectives. In *Intelligent Healthcare Systems*, ed. V. V. Estrela, 301–321. CRC Press.

[4] Estrela, V.V. and A. Deshpande (2023). Scale and resolution issues regarding medical images: Challenges ahead. In *Intelligent Healthcare Systems*, ed. V. V. Estrela, 90–109. CRC Press.

[5] Möller, D.P.F. and Haas, (2019). The connected car. In *Guide to Automotive Connectivity and Cybersecurity*. Prof. Dr. Dietmar P.F. Möller, Dr. Roland E. Haas., Springer International Publishing.

[6] Jumani, A.K. and Laghari, (2025). Image processing-based via agridrones for agriculture fields. In *Intelligent Designs, Innovations and Sustainability in Agriculture 4.0*, eds A. Deshpande, V.V. Estrela and A. Khan. CRC Press, pp. 214–241.

[7] Estrela, V.V., Jumani, A.K. and Laghari, (2023). Internet of Medical Things layers for medical cyber-physical systems. In *Intelligent Healthcare Systems*, ed. V.V. Estrela, 45–65. CRC Press.

[8] Belafonte-Young, D. (2018). 2018 Communication Technology Update and Fundamentals: Automotive Telematics. Broadcast Education Association (BEA) On-Location, Houston, Texas, United States. http://spiral.lynn.edu/facpubs/640

[9] Estrela, V.V. and Deshpande, (2023). Ad hoc networks in healthcare intelligent transportation systems humanitarian relief: MANETs, VANETs, and FANETs. In *Intelligent Healthcare Systems*, ed. V.V. Estrela, 66–89. CRC Press.

[10] Zhu, L., et al. (2019). Big data analytics in intelligent transportation systems: A survey. *IEEE Transactions on Intelligent Transportation Systems*, *20*, 383–398.

[11] Kulkarni, D.S. and Rodd, S.F. (2018).Extensive study of text-based methods for opinion mining. *IEEE International Conference on Inventive Systems and Control (ICISC)*, 523–527)

[12] Fathi, F., Abghour, N. and Ouzzif, M. (2018). From big data to better behavior in self-driving cars. In *Proceedings of the 2nd International Conference on Cloud and Big Data Computing*, 42–46.

[13] Kour, R., et al. (2019). Big data analytics for maintaining transportation systems. *Asset Analytics*.

[14] Nasr Azadani, M. and Boukerche, A.F.M. (2021). Driving behavior analysis guidelines for intelligent transportation systems. *IEEE Transactions on Intelligent Transportation Systems*, *23*, 6027–6045.

[15] Liu, J., et al. (2021). A survey of recent advances in driving behavior analysis. In *Proceedings of the 3rd International Symposium on Smart and Healthy Cities*, 145–157.

[16] Kargupta, H. and Kargupta, P. (2019). Analyzing driving data using the ADAPT distributed analytics platform for connected vehicles. *IEEE International Conference on Data Science and Advanced Analytics*, 590–599.

[17] Bouhoute, A. (2019). Advanced driving behavior analytics for an improved safety assessment and driver fingerprinting. *IEEE Transactions on Intelligent Transportation Systems*, *20*, 2171–2184.

[18] Nybacka, M., et al. (2014). Links between subjective assessments and objective metrics for steering, and evaluation of driver ratings. *Vehicle System Dynamics*, *52*(Suppl. 1), 31–50.

[19] Voß, G.M.I., Keck, C.M. and Schwalm, M. (2018). Investigation of drivers' thresholds of a subjectively accepted driving performance with a focus on automated driving. *Transportation Research Part F: Traffic Psychology and Behaviour*, *56*, 280–292.

[20] Chai, W., et al. (2024). Rethinking the evaluation of driver behavior analysis approaches. *IEEE Transactions on Intelligent Transportation Systems*, *25*(8), 9958–9966.

[21] Haghshenas, S.S., et al. (2024). Artificial intelligence-powered driver behavior analysis for fuel consumption optimization: A pathway to greener roads. In *Proceedings of the 10th International Conference on Control, Decision and Information Technologies*, 116–122.

[22] Milardo, Sebastiano & Rathore, Punit & Santi, Paolo & Ratti, Carlo. (2022). A Data-Driven Framework for Driving Style Classification. 10.1007/978-3-031-22137-8_19.

[23] Moshfeghi, Sonia & Jang, Jinwoo. (2025). Pattern Mining of Older Drivers' Driving Behavior Through Telematics-Data-Driven Unsupervised Learning. IEEE Sensors Journal. pp. 1–1. 10.1109/JSEN.2025.3592817.

[24] Abdelrahman, A.E., Hassanein, H.S. and Abuali, N.A. (2022). Robust data-driven framework for driver behavior profiling using supervised machine learning. *IEEE Transactions on Intelligent Transportation Systems*, *23*, 3336–3350.

[25] Krontiris, Ioannis & Grammenou, Kalliroi & Terzidou, Kalliopi & Zacharopoulou, Marina & Tsikintikou, Marina & Baladima, Foteini & Sakellari, Chrysi & Kaouras, Konstantinos. (2020). Autonomous Vehicles: Data Protection and Ethical Considerations. 1-10. 10.1145/3385958.3430481.

[26] Oueida, S., et al. (2021). A fair and ethical healthcare artificial intelligence system for monitoring driver behavior and preventing road accidents. *arXiv*, 2107.

[27] Stahl, B.C. and Wright, D. (2018). Ethics and privacy in AI and big data: Implementing responsible research and innovation. *IEEE Security & Privacy*, *16*, 26–33.

[28] Kumar, Amaresh. (2024). Exploring Ethical Considerations in AI-driven Autonomous Vehicles: Balancing Safety and Privacy. Journal of Artificial Intelligence General science (JAIGS) ISSN:3006-4023. 2. 125–138. 10.60087/jaigs.v2i1.p138.

[29] Bhattacharya, S., et al. (2025). AI revolution in insurance: Bridging research and reality. *Frontiers in Artificial Intelligence*, p. 8.

[30] Sun, S., et al. (2020). Assessing driving risk using Internet of Vehicles data: An analysis based on generalized linear models. *Sensors*, *20*.

[31] Hanafy, Mohamed and Ming, Ruixing. (2021). Machine Learning Approaches for Auto Insurance Big Data. Risks. 9. 42. 10.3390/risks9020042.

[32] Edymond, H. (2024). Advanced Data Science Applications in Vehicles: A Comprehensive Review. *International Journal of Technology and Systems*, *9*(1), 28–34. https://doi.org/10.47604/ijts.2402

[33] Driessen, Tom & Siebinga, Olger & de Boer, Thomas & Dodou, Dimitra & Waard, Dick & de Winter, Joost. (2024). How AI from Automated Driving Systems Can Contribute to the Assessment of Human Driving Behavior. Robotics. 13. 169. 10.3390/robotics13120169.

[34] Atakishiyev, S., et al. (2021). Explainable artificial intelligence for autonomous driving: A comprehensive overview and field guide for future research directions. *IEEE Access*, *12*, 101603–101625.

[35] Yun, Keon & Yun, Heesun & Lee, Sangmin & Oh, Jinhyeok & Kim, Minchul & Lim, Myongcheol & Lee, Juntaek & Kim, Chanmin & Seo, Jiwon & Choi, Jinyoung. (2024). A Study on Machine Learning-Enhanced Roadside Unit-Based Detection of Abnormal Driving in Autonomous Vehicles. Electronics. 13. 288. 10.3390/electronics13020288.

[36] Razmjooy, N., Deshpande, A., Khalilpour, M., Estrela, V.V., Padilha, R. and Monteiro, A.C.B. (2021). Optimal bidding strategy for power market based on improved world cup optimization algorithm. In *Metaheuristics and Optimization in Computer and Electrical Engineering*, Vol. 696, eds N. Razmjooy, M. Ashourian, & Z. Foroozandeh. Springer, ISBN: 978-3-030-56688-3

[37] Tong, J. and Ren, T. (2024). Research on multimodal intelligent navigation and AI+AR display design theory. *Journal of Research in Science and Engineering*, *6*(8), pp. 38–41.

[38] de Almeida, M.B. (2023). AI-based framework for automatically extracting high–low features from NDS data to understand driver behavior. https://doi.org/10.32469/10355/94205

[39] Farfan, W.S. and Saotome, (2020). Integrated optical flow for situation awareness, detection and avoidance systems in UAV systems. In *Imaging and Sensing for Unmanned Aircraft Systems*, Vol. 1, eds V.V. Estrela, J. Hemanth, O. Saotome, G. Nikolakopoulos and R. Sabatini, 47–74. IET.

[40] Mehta, A., et al. (2023). Securing the future: A comprehensive review of security challenges and solutions in advanced driver assistance systems. *IEEE Access*, *12*, 643–678..

[41] Kapilan, K., Bandara, S. and Dammalage, T. (2020). Vehicle accident detection and warning system for Sri Lanka using GNSS technology. In *Proceedings of the International Conference on Image Processing and Robotics*, 1–5.
[42] Yaqoob, S., et al. (2024). AI-driven driver behavior assessment through vehicle and health monitoring for safe driving—A survey. *IEEE Access*, *12*, 48044–48056.
[43] Laghari, A. A., et al. (2024). Quality of experience assessment in virtual/augmented reality serious games for healthcare: A systematic literature review. *Technology and Disability*, *36*, 17–28.
[44] Jumani, A.K., Shi, J., Laghari, A.A., Estrela, V.V., Sampedro, G.A., Almadhor, A., Kryvinska, N. and ul Nabi, A. (2024). Quality of experience that matters in gaming graphics: How to blend image processing and virtual reality. *Electronics*, *13*(15), 2998.
[45] Kondapaka, K.K. (n.d.). Integrating AI and augmented reality for enhanced driver assistance and navigation in modern vehicles. *Journal of Machine Learning for Healthcare Decision Support*. 4(1) pp. 151–188.
[46] Guirrou, H., Es-sadek, M. and Taher, Y. (2023). Artificial intelligence in self-driving: Study of advanced current applications. *International Journal of Engineering Trends and Technology*, *71*(8), 225–242.
[47] Deshpande, Anand and Estrela, Vania and Patavardhan, Prashant and Kallimani, G. (2023). Super-resolution in a world of scarce resources for medical imaging applications. In *Edge-AI in Healthcare: Trends and Future Perspectives*, eds S. Vyas et al.. CRC Press, pp. 183–196.
[48] Musicant, O., et al. (2017). Integrating kinematic- and vision-based information to better understand driving behaviour. In *Proceedings of the International Driving Symposium on Human Factors in Driver Assessment, Training and Vehicle Design*, *6*, 518–524.
[49] Chanmas, G., et al. (2024). Driving assistant using generative AI pre-generated messages in simulator-based driving assessment: A step towards low-cost simulator-based driving assessment. *Heliyon*, *10*(16), p. e35941.
[50] Anurag Sharma; Harsh Jangid; Raghvendra Singh Nirwan; Sohit Negi; Rahul Sharma. “Autonomous Vehicles: Challenges and Advancements in AI-Based Navigation Systems.” Volume. 8 Issue. 12, December - 2023 International Journal of Innovative Science and Research Technology (IJISRT), www.ijisrt.com. ISSN - 2456-2165, pp. 992–1003.https://doi.org/10.5281/zenodo.10427687
[51] Capitaine, M. and Cárdenas, H. (2023). Artificial intelligence and advanced driver assistance systems absorption in Mexico. *Ciencia Nicolaita*, *88*.
[52] Baccari, S., et al. (2024). Anomaly detection in connected and autonomous vehicles: A survey, analysis, and research challenges. *IEEE Access*, *12*, 19250–19276.
[53] Langner, L., et al. (2019). A driving scenario representation for scalable real-data analytics with neural networks. In *Proceedings of the IEEE Intelligent Vehicles Symposium*, 2215–2222.
[54] Shih, Y.-T. and Chen, P.-C. (2013). *System and Method of Reducing Noise* (Patent).
[55] Amiri, M. and Jensen, R. (2016). Missing data imputation using fuzzy-rough methods. *Neurocomputing*, *205*, 152–164.
[56] Hoque, A. and Ahmed, (2025). AI-assisted imaging and navigation in minimally invasive cardiac interventions: A systematic review. *Insights—Journal of Health and Rehabilitation*. Vol. 3. pp. 538–544.
[57] Chougule, A. and Chamola, (2023). A comprehensive review on limitations of autonomous driving and its impact on accidents and collisions. *IEEE Open Journal of Vehicular Technology*, Vol. 5, pp. 1–20.
[58] Vedantham, A.K. (2024). Augmented reality in modern vehicles: Enhancing safety and driver experience.*International Journal for Multidisciplinary Research*, *6*(6).
[59] Yamaguchi, Y. (2024). *AR-Navi: An in-vehicle navigation system using video-based augmented reality technology, 2024*. pp. 1139–1147.

[60] Cao, J., et al. (2010). Analysis and application of the optimal path based on ArcGIS geodatabase network model. In *Environmental Science and Information Application Technology*, IEEE. Vol. 1, 434–437.

[61] Halim, Z., Kalsoom, R. and Baig, A.R. (2016). Profiling drivers based on driver-dependent vehicle driving features. *Applied Intelligence*, *44*(3), 645–664.

[62] Hauer, Ezra. (2015). *The Art of Regression Modeling in Road Safety*. 10.1007/978-3-319-12529-9.

[63] Yeo, M.V.M., et al. (2009). Can SVM be used for automatic EEG detection of drowsiness during car driving? *Safety Science*, *47*(1), 115–124.

[64] Oguchi, K. and Weir, D.F.R. (2008). *System for predicting driver behavior* (US Patent No. US20080167820A1).

[65] Fu, Z., Hu, W. and Tan, T. (2005). Similarity-based vehicle trajectory clustering and anomaly detection. In *Proceedings of the International Conference on Image Processing*, Vol. 2, 602–605.

6 Smart Vision: Powering Vehicle Detection with Deep Learning

Gautam Arjun Dematti[1*], Uttam Patil[2], Kuldeep Sambrekar[3], Karuna C. Gull[4] and Anjanabhargavi Kulkarni[5]

The rapid growth of urbanization and the escalating volume of vehicular traffic have underscored the urgent need for efficient and accurate vehicle detection systems as part of modern intelligent transportation infrastructures. In this study, a deep learning rooted in Convolutional Neural Networks (CNNs)–based approach was employed to detect and classify Indian road vehicles using the YOLOv10n model—one of the latest and most lightweight architectures optimized for real-time object detection. The model was trained and evaluated using a curated dataset comprising five major vehicle categories commonly found on Indian roads: Truck, Car, Bus, Tempo Traveller, and Tractor. To ensure compatibility with the model's input structure, image pre-processing steps such as resizing and normalization were applied. A pre-trained YOLOv10n model was fine-tuned using transfer learning techniques to adapt to the specific visual characteristics of Indian vehicles and traffic scenes. Performance evaluation revealed that the YOLOv10n model achieved strong detection results, particularly for Cars and Buses, with mean Average Precision (mAP) scores of 84.55% and 82.49%, respectively. The detection of Tempo Travellers and Tractors also yielded satisfactory outcomes, although minor inaccuracies were observed in scenarios involving occlusion or visually complex backgrounds. To provide a comparative benchmark, an R-CNN model was also trained and tested on the same dataset. While R-CNN achieved acceptable accuracy, it lagged behind YOLOv10n across most performance metrics, particularly in inference speed and real-time applicability. Overall, the findings underscore YOLOv10n's effectiveness and suitability for practical deployment in Indian traffic environments, where rapid, accurate detection of varied vehicle types is essential for data-driven traffic management solutions.

1 Department of Computer Science and Engineering, Angadi Institute of Technology and Management, Belagavi, Karnataka, India. (Affiliated to Visvesvaraya Technological University, Belagavi, Karnataka, India.)

2 Department of Computer Science and Engineering, Jain College of Engineering, Belagavi, Karnataka, India.

3 Department of Computer Science and Engineering (AI & ML), KLS Gogte Institute of Technology, Belagavi, Karnataka, India.

4 Computer Science and Business Systems Department, S.G. Balekundri Institute of Technology, Visvesvaraya Technological University, Belagavi, Karnataka, India.

5 Department of Computer Science and Engineering, Visvesvaraya Technological University, Belagavi, Karnataka, India.

* Corresponding author: dematti.gautam@gmail.com

DOI: 10.1201/9781003648147-06

1. Introduction

Intelligent Transportation Systems (ITS) is a broad, interdisciplinary area that applies information, communication, and control technologies to transportation infrastructure and vehicles. ITS improves transportation safety, efficiency, and sustainability, boosting mobility and quality of life for all road users. ITS includes advanced traffic management, traveler information systems, public transportation, and vehicle control. Accurate, real-time, and complete traffic data are essential to any efficient ITS. Computer vision (CV) offers several effective solutions for these situations [1–10]. These CV-produced data underpin wise judgments, strategies, and transportation system optimization. In this context, vehicle counting and classification are crucial data elements. These measurements are essential to practically every aspect of a healthy ITS ecosystem. Consider the real-time ITS subcomponents [11]:

- *Advanced Traffic Management Systems (ATMS)*: This is where vehicle counting and classification truly shine. ATMS rely on precise data to dynamically adjust traffic signal timings, manage ramp meters on highways, re-route traffic during congestion or incidents, and optimize lane usage. For instance, prioritizing emergency vehicles or adapting signal timings based on the proportion of heavy vehicles requires this granular data [12, 13].
- *Advanced Traveler Information Systems (ATIS)*: Real-time traffic information, such as current road speeds, travel times, and incident locations, is disseminated to travelers through various channels (e.g., navigation apps, roadside message boards, radio). This information is derived directly from aggregated vehicle counts and speeds obtained through effective monitoring. Accurate classification data can also inform travelers about specific lane restrictions or conditions pertinent to different vehicle types [14–17].
- *Traffic and Incident Management*: Rapid detection and verification of incidents are critical for minimizing their impact. Vision-based systems, underpinned by vehicle detection and tracking, can automatically flag unusual traffic behavior, alerting operators for immediate response.
- *Urban and Transportation Planning*: For long-term strategic planning, historical vehicle count and classification data are invaluable. Urban planners use this information to understand peak-hour demands, forecast future traffic volumes, assess the impact of new developments, and make informed decisions regarding infrastructure expansion, public transport investments, and road network modifications.
- *Safety and Enforcement*: Vehicle counting and classification data contribute to identifying hazardous road segments. They can also support automated enforcement by identifying speed violations or red-light running, and by classifying vehicles to apply different regulations or penalties.
- *Environmental Sustainability*: Traffic congestion significantly contributes to air pollution and greenhouse gas emissions. By optimizing traffic flow through precise monitoring and management, ITS, powered by vehicle analytics, helps

reduce idling times and promotes smoother traffic, directly contributing to environmental sustainability goals.

Vehicle detection is an essential component that plays a crucial role across domains such as traffic management, road safety, and smart city development. It facilitates the monitoring of vehicles on roadways, the management of traffic flow, and the identification of traffic violations [18, 19]. As time has progressed, various approaches have been proposed to recognize vehicles from images and videos, with deep learning and neural networks significantly enhancing the accuracy and efficiency of these tasks. Among the numerous object detection frameworks available, the YOLO (You Only Look Once) series has gained considerable prominence for its fast and accurate detection.

Despite its importance, vehicle object detection faces several challenges. The large variety of vehicle shapes, sizes, and types in countries like India creates complexity in training detection models. Environmental factors such as poor lighting, shadows, rain, and fog degrade image quality, making detection difficult. Additionally, occlusions caused by overlapping vehicles or roadside objects often result in missed detections or false identifications [20, 21, 22]. These challenges require robust algorithms capable of handling diverse real-world scenarios to ensure reliable performance. Traditional vehicle detection techniques like Histogram of Oriented Gradients (HOG) [11, 23], Haar-like features [13], and Local Binary Pattern (LBP) rely on rigid, handcrafted features, i.e., manually designed rules to find specific patterns like edges or textures. Their main limitation is a lack of robustness in real-world conditions. They are susceptible to variations in vehicle viewpoint, scale, lighting, and weather, leading to poor accuracy. They also struggle significantly with occlusion and cluttered backgrounds, leading to high false-positive and false-negative rates.

The emergence of modern vision-based systems was a direct response to the failures of older handcrafted techniques [24, 25, 26]. This made them brittle and unreliable in real-world traffic, where vehicle appearance changes constantly due to different viewpoints, lighting, weather, and partial occlusion. Modern systems, powered by deep learning via Convolutional Neural Networks (CNNs), represent a paradigm shift. Instead of being programmed with fixed rules, they learn the essential features of a vehicle from vast datasets.

This research has two main contributions: (i) the use of the YOLOv10n model for vehicle detection and (ii) the advent of the Indian Vehicle Dataset benchmark. Leveraging this dataset allows the model to learn features and patterns specific to Indian vehicles and traffic conditions.

YOLOv10n is a lightweight model specifically designed to perform effectively on devices with constrained computational resources while maintaining robust detection performance. YOLOv10n was chosen because it balanced speed, compactness, and suitability for real-time deployment. Its design enables operation on platforms with constrained computational capabilities. The objective of this study is to demonstrate that the YOLOv10n model, when fine-tuned on the Indian Vehicle Dataset, can achieve accurate vehicle detection. Such performance is expected to facilitate the advancement of intelligent traffic monitoring and management systems tailored for Indian road conditions. YOLO significantly reduces the architectural complexity and the number of components to manage.

Popular datasets like COCO have been widely used to train and evaluate object detection models; however, they generally lack Indian vehicle types and traffic conditions. In congested and complex traffic, scooters, motorbikes, rickshaws, buses, lorries, and cars share Indian roads. Images from across India were used to create the Indian Vehicle Dataset. The work in [27] uses a lightweight YOLOv10n model fine-tuned on an Indian Vehicle Dataset to address impediments in Indian traffic. This research aims to improve Indian transportation. The results compare the performance of a YOLOv10n and a region-based convolutional neural network (R-CNN). The YOLOv10n architecture offers a significant advantage in implementation simplicity compared to the original R-CNN architecture. This ease of implementation stems from YOLO's unified, single-stage approach to object detection in contrast to R-CNN's complex, multi-stage pipeline.

Section II provides an overview of Related Work. Section III elaborates traditional methodologies. Section IV elaborates the proposed methodology. Section V provides analysis of results and Section VI concludes the work.

2. Related Work

In [1], while traditional traffic monitoring systems have relied on embedded sensors [28], Parthasarathy et al.'s innovative approach leverages Python's OpenCV library to implement a CV pipeline that addresses two critical urban mobility challenges: real-time differentiation between motorcycles and light vehicles, and dynamic traffic density estimation through centroid-based tracking. Their methodology combines adaptive background subtraction with morphological segmentation techniques, achieving 92.4% classification accuracy under daylight conditions. In [2], implementation of the YOLOv8 architecture advances vision-based traffic analytics through dual-function detection, achieving 96% counting accuracy while simultaneously estimating vehicle speeds. In [3], implementation demonstrates YOLOv8's efficacy in real-time traffic analysis through a specialized four-class detection framework. In [4], a CNN-based pipeline is developed achieving 94.1% vehicle classification accuracy via optimized Region of Interests (ROIs) and MobileNetV3 backbone, processing 38 Frames Per Second (FPS) on edge devices. In [5] there is deployed a Single-Shot Detector (SSD)-MobileNet with TensorFlow for multi-task vehicle analysis: counting (92% accuracy).

In [6], boost classification accuracy via a residual network with attention mechanisms and focal loss, optimizing feature extraction for precise vehicle identification and incorporating attention-guided feature refinement and focal loss to mitigate class imbalance, achieves 94.3% accuracy across diverse vehicle types. In [7], Oriented FAST and Rotated BRIEF (ORB) feature detection, frame differencing, and sliding windows are combined for embedded vehicle counting. It features low-resource ORB for real-time processing, frame differencing for motion detection, and embedded vision optimization. In [8], early systems employed background subtraction/edge detection (pre-2010), while post-2010 YOLOv5 advancements enabled real-time detection (96% accuracy, 30+ FPS) via deep learning. In [9], a virtual detection zone system using YOLOv3 for five-class vehicle classification is implemented with Kalman-filter-enhanced tracking via custom sorting algorithms. In [10], an Inception model-driven system for vehicle counting/classification is implemented, achieving a 70.75% true positive rate at 5 FPS inference speed. Early systems relied on manually designed features, such as edges and textures, often using algorithms like HOG [11] and Haar Cascades [14]. Architectures such as YOLO and SSD [28] have enabled real-time processing of video feeds, delivering far richer outputs ranging from object classification and tracking to full-scene semantic analysis [30]. Occlusion, where one vehicle blocks another from the camera's view, can disrupt tracking continuity and reduce counting accuracy [20, 21 22]. Moreover, variation in scale and perspective, where vehicle appearance shifts based on distance from the camera, demands sophisticated multi-scale modeling to maintain consistency.

Real-time processing presents another technical hurdle. High-resolution video analysis requires substantial computational resources, often necessitating Graphics Processing Units (GPUs) or specialized edge devices [31, 32, 33].

3. Vehicle Detection, Counting, and Classification Techniques

Vehicle detection is the foundational task in vision-based traffic analysis, involving the identification and localization of vehicles within an image or video frame. This typically results in a bounding box drawn around each detected vehicle, along with an associated confidence score. The accuracy and speed of this initial detection step profoundly impact the performance of subsequent processes like tracking, counting, and classification.

3.1 Foundations and Progression of Vehicle Detection in Visual Traffic Systems

Before deep learning, vehicle detection systems relied on traditional machine learning pipelines that incorporated manually crafted features. Background subtraction techniques were often used to isolate moving objects from a static scene. These candidate regions were then analyzed using feature descriptors, such as HOG [11], Scale-Invariant Feature Transform (SIFT) [12], or Haar-like features [13].

While the traditional methods introduced foundational strategies for automated detection, they were significantly limited in scope and adaptability. The reliance on handcrafted features meant that systems lacked flexibility across varying conditions and required considerable expert input. Performance degraded rapidly under difficult lighting, heavy occlusion, or complex backgrounds. Moreover, the computational cost of scanning high-resolution images using sliding windows made real-time applications impractical. These shortcomings highlighted the need for more adaptable, scalable approaches, which led to the rise of deep learning techniques [34, 35, 36]. The introduction of CNNs marked a pivotal shift in visual data analysis. CNNs enabled learning and extraction of relevant features directly from raw image pixels, thereby eliminating the need for manual feature design [34, 35, 36].

Modern object detection architectures integrate these CNN components within larger frameworks optimized for both speed and accuracy. Networks such as YOLO and SSD [18, 29] exemplify this progress by performing detection in a single pass, significantly reducing inference time and enabling real-time analysis.

Modern vehicle detection systems have significantly evolved with the advent of deep learning, transitioning from traditional image analysis to end-to-end frameworks capable of precise object recognition, classification, and counting in real time. Among the core architectures, two major paradigms have emerged: two-stage and one-stage object detectors.

- *Two-stage detectors*, such as the R-CNN family [37, 38], operate by first generating candidate regions likely to contain objects and then performing classification and bounding box refinement on these regions. The original R-CNN approach, although groundbreaking, was computationally intensive due to its reliance on external region proposal algorithms and repetitive CNN operations.
- *One-stage detectors* streamline detection by predicting bounding boxes and class probabilities in a single pass. This architectural simplicity allows for much faster inference, making these models ideal for real-time tasks such as traffic monitoring. The YOLO family exemplifies this category.

Other one-stage models include SSD [18, 29], which makes predictions from multiple layers within the network to detect objects of different scales. It employs a variety of default anchor boxes across feature maps, balancing speed and accuracy.

Several key components underpin the performance of modern detectors. Anchor boxes serve as reference bounding boxes, simplifying the prediction task by learning offsets rather than absolute coordinates. Non-maximum Suppression (NMS) filters overlapping boxes to retain the most confident detections [39], and Intersection over Union (IoU) [40] is used to assess the accuracy of predictions by comparing the overlap between predicted and actual bounding boxes. Despite the remarkable advancements in vision-based systems for vehicle counting and classification, particularly driven by deep learning, the field continues to grapple with significant challenges. These limitations often arise from the inherent complexities of real-world traffic environments and the demand for robust, reliable performance under diverse conditions. Simultaneously, active research is pushing the boundaries, exploring novel approaches and emerging technologies that promise to address current hurdles and unlock new capabilities for intelligent transportation [44, 45, 46, 47].

4. Methodology

4.1 YOLOv10n Architecture

The YOLO family of object detectors treats object detection as a single-stage regression problem: an image is divided into a grid, and each grid cell predicts bounding boxes [41, 42, 43]. YOLO's single-shot design allows for faster inference speeds. YOLOv10n follows a "neural architecture search" and "manual pruning" approach to lower the quantity of parameters and floating-point operations (FLOPs). The architecture integrates improvements from previous YOLO versions while removing redundant components, such as overly deep convolutional layers and computationally expensive attention blocks. The "n" in YOLOv10n refers to "nano," indicating that this model is the smallest and most efficient variant in the YOLOv10 family. YOLOv10n consists of fewer residual blocks and smaller kernel sizes to reduce computational complexity.

The detection head in YOLOv10n is intended to minimize complexity. It consists of a minimal convolutional layer and uses anchor-free prediction heads. Anchor-free detection eliminates the need for pre-defined anchor boxes and instead predicts object center points and sizes directly. This reduces computational requirements and simplifies training. In anchor-free methods, the loss for bounding box regression is generally computed using the Generalized Intersection over Union (GIoU) or Complete IoU Complete Intersection over Union (CIoU):

$$\mathrm{Loss}_{\mathrm{CIoU}} = 1 - \mathrm{IoU} + \frac{\rho^2(\mathbf{b}, \mathbf{b}^{gt})}{c^2} + \alpha v \qquad \text{...(6.1)}$$

where:

- ρ is the Euclidean distance between the centers of predicted and ground-truth boxes,
- c is the diagonal length of the smallest enclosing box,
- v measures aspect ratio consistency, and
- α is a trade-off factor.

This loss function helps the model to learn accurate bounding boxes without relying on heuristically chosen anchor sizes. YOLOv10n is typically trained with stochastic gradient descent or Adam optimizer with cosine learning rate scheduling for stable convergence.

4.2 Proposed Methodology

To evaluate the effectiveness of the object detection model, a structured training and testing procedure was implemented using the curated Indian Vehicle Dataset. This dataset was initially divided into two subsets: a training set for model learning and a testing set for performance evaluation. Each image was annotated with bounding boxes and vehicle class labels, ensuring high-quality ground truth data. The division allowed the model to be trained on a broad representation of vehicle types and environmental conditions, while reserving unseen data for unbiased evaluation.

Pre-processing was conducted to ensure consistency and compatibility with the chosen detection model. All images were resized to match the YOLOv10n architecture's input dimensions and normalized to standardize pixel values, thereby enhancing learning stability and convergence during training. A *pre-trained YOLOv10n model* was selected due to its efficiency and suitability for edge deployment, as well as its strong real-time performance. Leveraging transfer learning, the model was fine-tuned on the Indian dataset to adapt to regional vehicle appearances and traffic characteristics. This approach allowed the model to retain general object detection knowledge while learning the specific visual patterns associated with local vehicle categories such as Truck, Car, Bus, Tempo Traveller, and Tractor.

During the *training phase*, the model's performance was systematically evaluated using key metrics: Precision, Recall, F1 Score, and mean Average Precision (mAP@[0.5:0.95]). These metrics provided a comprehensive understanding of the model's ability to accurately detect and localize different vehicle types under diverse real-world conditions.

Accuracy, Precision, Recall, and F1 Score are key metrics for evaluating classification models. Accuracy measures the overall correctness of predictions, calculated as the ratio of correctly predicted samples to total samples. Precision indicates how many of the predicted positives are actually positive, focusing on minimizing false positives. Recall (or sensitivity) measures how many actual positives were correctly identified, emphasizing the reduction of false negatives. F1 Score is the harmonic mean of Precision and Recall, providing a balanced measure when both false positives and false negatives are important, especially useful for imbalanced datasets.

Accuracy
$$\text{Accuracy} = \frac{TP + TN}{TP + TN + FP + FN} \quad ...(6.2)$$

Precision
$$\text{Precision} = \frac{TP}{TP + FP} \quad ...(6.3)$$

Recall
$$\text{Recall} = \frac{TP}{TP + FN} \quad ...(6.4)$$

F1 Score
$$\text{F1} = 2 \cdot \frac{\text{Precision} \cdot \text{Recall}}{\text{Precision} + \text{Recall}} \quad ...(6.5)$$

5. Result Analysis

5.1 Indian Vehicle Dataset

The Indian Vehicle Dataset utilized in this study was meticulously curated to reflect the diverse, dynamic, and often challenging traffic scenarios characteristic of

Indian roads. Recognizing the unique nature of Indian traffic—marked by varying vehicle types, road conditions, and environmental influences—the dataset was compiled using images sourced from both urban centers and rural regions across the country. These images capture a broad spectrum of real-world conditions, including different times of day, weather variations, and a variety of traffic densities and road backgrounds. The dataset focuses on five key vehicle categories that are commonly found in Indian transportation systems: Truck, Car, Bus, Tempo Traveller, and Tractor. These categories were selected for their frequent occurrence on Indian roads and their importance to traffic flow and infrastructure planning. Each image in the dataset is annotated with precise bounding boxes delineating the positions of each vehicle, along with accurate class labels.

5.2 Results

The performance evaluation of the YOLOv10n object detection model was conducted using the Indian Vehicle Dataset which comprised of five major vehicle categories: Truck, Car, Bus, Tempo Traveller, and Tractor. In Table 6.1, for the Truck category, a Precision of 75.62%, a Recall of 72.34%, and an F1 Score of 73.94% were achieved. The mAP@[0.5:0.95] was recorded at 70.31%. This indicated that trucks were identified with a moderate degree of accuracy and consistency in bounding box placement. Although a reasonable number of actual trucks were correctly identified, the localization accuracy was relatively lower.

With regard to Car detection, the highest Precision of 84.26% was observed among all categories. However, the Recall was lower at 70.87%, leading to an F1 Score of 76.99%. The highest mAP@[0.5:0.95] value of 84.55% was noted for this category, indicating highly accurate localization. These findings imply that false positives were fewer, but some actual car instances were missed. For the Bus category, a well-rounded performance was achieved with a Precision of 80.98% and a Recall of 82.99%, yielding the highest F1 Score of 81.97%. The mAP@[0.5:0.95] was also high at 82.49%, indicating that buses were detected effectively, with both high confidence and accuracy.

The Tempo Traveller category showed consistent values with a Precision of 78.98% and a Recall of 79.02%, producing an F1 Score of 79.00%. The mAP@

Table 6.1 Results for Yolov10n for 70:30 split

Category	*Precision (%)*	*Recall (%)*	*F1 Score (%)*	*mAP@[0.5:0.95] (%)*
Truck	75.62	72.34	73.94	70.31
Car	84.26	70.87	76.99	84.55
Bus	80.98	82.99	81.97	82.49
Tempo Traveller	78.98	79.02	79.00	73.19
Tractor	72.34	80.62	76.26	72.73

[0.5:0.95] was 73.19%, indicating strong performance in detection and localization. Minor inaccuracies in bounding box placement may have affected the mAP score, possibly due to background complexity or partial occlusion. In the case of Tractor detection, a lower Precision of 72.34% was observed, while a higher Recall of 80.62% was achieved. This resulted in an F1 Score of 76.26%, with a mAP@[0.5:0.95] of 72.73%. The lower Precision indicated that more false positives were produced for this category.

Table 6.2 shows the results of applying the Indian Vehicle Dataset to the R-CNN. In Table 6.2, for the Truck category, Precision was 71.12% and Recall was 70.11%. This resulted in an F1 Score of 71.58%. The mean Average Precision (mAP@[0.5:0.95]) was 70.61%. These values indicate that trucks were detected with a moderate degree of accuracy and consistency. In the case of Cars, a higher precision of 74.26% was observed, while the Recall stood at 70.76%. The F1 Score was calculated as 72.19%, and the mAP@[0.5:0.95] was recorded at 74.15%. These findings imply that the car detection was handled with relatively high accuracy and reliable localization.

For the Bus category, a precision of 70.48% and a Recall of 72.99% were obtained. This led to an F1 Score of 71.83%. The mAP@[0.5:0.95] for this category was 72.49%, indicating balanced performance in detecting and localizing buses. In the case of Tempo Travellers, Precision was measured at 65.18% and Recall was higher at 75.93%. As a result, the F1 Score reached 72.70%, while the mAP@[0.5:0.95] was recorded at 68.59%. These values indicate that Tempo Travellers were identified with good Recall but slightly lower localization accuracy. For the Tractor category, a precision of 70.31% and a recall of 68.12% were observed. The F1 Score was calculated as 70.12%, and the mAP@[0.5:0.95] stood at 72.23%. This suggests that tractors were detected with reasonable accuracy, though some variation in detection consistency may have occurred.

Table 6.2 Results for R-CNN for 70:30 split

Category	*Precision (%)*	*Recall (%)*	*F1 Score (%)*	*mAP@[0.5:0.95] (%)*
Truck	71.12	70.11	71.58	70.61
Car	74.26	70.76	72.19	74.15
Bus	70.48	72.99	71.83	72.49
Tempo Traveller	65.18	75.93	72.70	68.59
Tractor	70.31	68.12	70.12	72.23

6. Conclusion

Based on the evaluation results, the YOLOv10n model was discovered to perform better overall regarding detection accuracy and localization across most vehicle categories. Higher values of Precision, Recall, and F1 Score were recorded for categories such as Car, Bus, and Tempo Traveller when compared to the R-CNN model. Notably, the Car category achieved the highest Precision (84.26%) and mAP@[0.5:0.95] (84.55%) with YOLOv10n, reflecting its strong ability to localize car instances accurately while minimizing false positives. Similarly, Bus detection showed a well-balanced performance, achieving the highest F1 Score of 81.97% and a high mAP of 82.49%. These findings imply that the YOLOv10n model offered more reliable and consistent detection across multiple classes, particularly in complex road environments. The R-CNN model showed slightly lower performance across all metrics in most categories. For example, Tempo Traveller detection under R-CNN resulted in a lower Precision of 65.18% and a lower mAP of 68.59%, compared to the 78.98% Precision and 73.19% mAP achieved by YOLOv10n. While R-CNN did manage a fairly balanced outcome in some categories, such as Truck and Tractor, the overall results indicated that YOLOv10n provided stronger localization accuracy and more stable detection across the dataset.

Future research may explore integrating temporal video data, multi-class tracking, and edge optimization to further enhance the model's relevance in real-time, resource-constrained environments.

References

[1] Parthasarathy, S., Narendira Kumar, V.K., Raj, P., Brintha, R.M., Amali, S.M.J. and Thangasankaran, R. (2024). Robust approach to vehicle detection and counting with centroid-based tracking. *2024 International Conference on IoT Based Control Networks and Intelligent Systems*, Bengaluru, 1172–1179.

[2] Mostafa, S.A., Gaber, A. and Shehata, M. (2024). Automated vehicle counting and speed estimation using YOLOV8 and computer vision. *Journal of Al-Azhar University Engineering Sector*, *19*(73), 1382–1395.

[3] Bashar, S. and Abdulamir, K. (2024). Vehicle classification and counting for traffic analysis based on single-stage YOLOv8 model. *Iraqi Journal of Computers, Communicatons, Control and Systems Engineering*, *24*(2), 41–53.

[4] Shamimullah, S. and Hanirex, D.K. (2024). Integrated image processing pipeline for improved vehicle identification and counting in video streams. In Proceedings of the 2024 International Conference on Advances in Computing, Communication and Applied Informatics (ACCAI) (pp. 1–7). IEEE.

[5] Anand, V. and Asst, V.R. (2024). Advanced traffic monitoring: Precise vehicle counting, speed estimation, color recognition, and type classification utilizing ROC curve analysis. In Proceedings of the IEEE International Conference on Signal Processing (ICCSP) (pp. 349–354). IEEE

[6] Alkhalaf, S., Alfarraj, O. and Alzubi, A.A. (2024). An innovative model for detecting vehicles based on machine vision. In Proceedings of the 2024 10th International Conference on Computing and Data Engineering (ICCDE '24) (pp. 28–34). Association for Computing Machinery.

[7] Wang, M., Zhao, X., Chen, Z. and Chen, Q. (2023). Vehicles detection and counting method from overhead view based on embedded vision. In 2023 IEEE 6th International Conference on Pattern Recognition and Artificial Intelligence (PRAI) (pp. 164–169). IEEE.

[8] Premaratne, P., Kadhim, I.J., Blacklidge, R. and Lee, M. (2023). Comprehensive review on vehicle detection, classification and counting on highways. *Neurocomputing, 556*, 126627.

[9] Anil, J.M., Mathews, L., Renji, R., Jose, R.M. and Thomas, S. (2023). Vehicle counting based on convolution neural network. In *Proceedings of the 7th International Conference on Intelligent Computing and Control Systems (ICICCS)*, Madurai, India, 695–699.

[10] Imam Rifai, M., Sudibyo, R.W., Kurniawan, A.D.S., Zen Samsono Hadi, M., Mahmudah, H. and Sa'Adah, N. (2022). Automatic vehicle classification and counting system using inception model. In *Proceedings of the 14th International Conference on Information Technology and Electrical Engineering (ICITEE)*, Yogyakarta, Indonesia, 275–279.

[11] Alanazi, F. and Alenezi, M. (2024). A framework for integrating intelligent transportation systems with smart city infrastructure. Journal of Infrastructure, *Policy and Development, 8*(5), 3558.

[12] S,M., Shetty, K., Kodipalli, A., Rao, T. and S,K. (2023). Emergency vs non-emergency vehicle classification: Enhancing intelligent traffic management systems. In 2023 International Conference on Network, Multimedia and Information Technology (NMITCON) (pp. 1–6). IEEE.

[13] Saha, C., Tran, T.H. and Syamal, S. (2024). Enhancing automotive safety through advanced object behaviour tracking for intelligent traffic and transport system. *2024 IEEE International Workshop on Metrology for Automotive (MetroAutomotive)*, 70–75.

[14] Heijden, R.W. (2018). Misbehavior detection in cooperative intelligent transport systems. Ausgezeichnete Informatikdissertationen.

[15] Swathi, T., Nousheen, T., Rohith, V., Rohith, T. and Abhyuday, T. (2025). Smartroad: AI-powered accident detection and traffic flow analysis for intelligent transportation monitoring. *Scientific Digest: Journal of Applied Engineering*, 12(1) pp. 472–477.

[16] Hemanth, J., Wang, L., Tavares, J.M.R.S., Shi, F. and Estrela, V.V. (2018). Editorial. Special Issue: *Soft Computing Techniques for Image Analysis in the Medical Industry – Current Trends, Challenges and Solutions, IEEE Access, 6*, 39487–39489.

[17] Joshi, A., Sharma, S.T. and Gentne, C. (2025). Ultrasonic-based transportation mode detection in urban environments. *2025 International Conference on Indoor Positioning and Indoor Navigation (IPIN)*, 1–6.

[18] Dahri, F.H., Laghari, A.A., Dahri, N.A., Abro, G., Estrela, V.V., de Oliveira, G.G., Iano, Y. and Chuma, E.L. (2024). Real-time detection of UAV small target's infrared imaging via OSTD-YOLOv8. *2024 IEEE Uruguayan Conference on Electrical Engineering, Electronics and Computer Science*, 1–5.

[19] Estrela, V.V., Deshpande, A., Stutz, D., de Assis, J.T., Jumani, A.K., Khan, A.A., Shi, F., Yin, S. and Lin, Yu-Da (2023). Ad hoc networks in healthcare intelligent transportation systems humanitary relief—MANETs, VANETs, and FANETs. In *Intelligent Healthcare Systems*, ed. V. V. Estrela, 66–89. CRC Press.

[20] Upadhyaya, V. and Tiwari, N. (2024). Monocular vision based approach for occlusion detection and handling: A way forward for advanced driver assistance systems. *International Journal of Intelligent Transportation Systems Research, 22*, 205–215.

[21] Carmo, F.P., Estrela, V.V. and de Assis, J.T. (2009) Estimating motion with principal component regression strategies. In: *Proceedings of the 2009 IEEE International Workshop on Multimedia Signal Processing (MMSP'09)*, Rio de Janeiro, Brazil, IEEE, 1–6.

[22] Chiang, C., Jaber, M., Chai, K.K. and Loo, J. (2023). Distributed acoustic sensor systems for vehicle detection and classification. *IEEE Access, 11*, 31293–31303.

[23] Dalal, N. and Triggs, B. (2005). Histograms of oriented gradients for human detection. *IEEE Conference on Computer Vision and Pattern Recognition, 1*, 886–893.

[24] Monteiro, A.C.B., França, R.P. and Estrela, V.V. (2021) Metaheuristics applied to blood image analysis. In *Metaheuristics and Optimization in Computer and Electrical Engineering. Lecture*

Notes in Electrical Engineering, eds N. Razmjooy, M. Ashourian, & Z. Foroozandeh, vol. 696. Springer, Cham.

[25] Deshpande, A., Estrela, V.V., Patavardhan, P. and Kallimani, G. (2023) Super-resolution in a world of scarce resources for medical imaging applications. In *Edge-AI in Healthcare: Trends and Future Perspectives*, eds S. Vyas et al. CRC Press.

[26] Irhebhude, M.E., Kolawole, A.O. and Izuegbu, M.C. (2023). Handcrafted and transfer learned feature techniques for vehicle make and model recognition on Nigerian road. *Malaysian Journal of Applied Sciences,* 8(2), pp. 83–98.

[27] Hua, Z., Aranganadin, K., Yeh, C., Hai, X., Huang, C., Leung, T., Hsu, H., Lan, Y. and Lin, M. (2025). A benchmark review of YOLO algorithm developments for object detection. *IEEE Access*, *13*, 123515–123545.

[28] Helbing, D. (2001). Traffic and related self-driven many-particle systems. *Reviews of Modern Physics*, *73*(4), 1067–1141.

[29] Liu, W., Anguelov, D., Erhan, D., Szegedy, C., Reed, S., Fu, C.Y. and Berg, A.C. (2016). SSD: Single-shot multibox detector. In *Proceedings of the European Conference on Computer Vision*, 21–37. Springer, Cham.

[30] Girshick, R., Donahue, J., Darrell, T. and Malik, J. (2014). Rich feature hierarchies for accurate object detection and semantic segmentation. In *Proceedings of the IEEE Conference on Computer Vision and Pattern Recognition*, 580–587.

[31] Ojala, T., Pietikäinen, M. and Mäenpää, T. (2002). Multiresolution gray-scale and rotation invariant texture analysis with local binary patterns. *IEEE Transactions on Pattern Analysis and Machine Intelligence*, *24*(7), 971–987.

[32] Tan, M. and Le, Q.V. (2019). EfficientNet: Rethinking model scaling for convolutional neural networks. In *International Conference on Machine Learning*, 6105–6114. Proceedings of Machine Learning Research. (PMLR).

[33] Kang, S., Byun, S., Yoon, Y. and Park, Y. (2018). Study on single camera-based ANPR system for improvement of vehicle number plate recognition on multi-lane roads. *Industrial & Manufacturing Engineering eJournal*, 2(1), pp. 11–16.

[34] Farfan, W.S., Saotome, O., Estrela, V.V. and Razmjooy, N. (2020). Integrated optical flow for situation awareness, detection and avoidance systems in UAV systems. In *Imaging and Sensing for Unmanned Aircraft Systems*, eds V.V. Estrela, J. Hemanth, O. Saotome, G. Nikolakopoulos and R. Sabatini, *1*(3), 4774. Institution of Engineering and Technology (IET).

[35] Estrela, V.V., Bukhsh, K., Malook Rind, M., Shaikh, S., Khan, A.A., Laghari, A.A., Deshpande, A., Andreopoulos, N. and Terziev, A. (2023). Deep learning as a driving force for better drug development. In *Intelligent Healthcare Systems*, ed. V.V. Estrela, 217–235. CRC Press.

[36] Deshpande, A., Patavardhan, P., Estrela, V.V. and Razmjooy, N. (2020). Deep learning as an alternative to super-resolution imaging in UAV systems. In *Imaging and Sensing for Unmanned Aircraft Systems*, Vol. 2, eds V.V. Estrela, J. Hemanth, O. Saotome, G. Nikolakopoulos and R. Sabatini, 177–212. IET.

[37] Girshick, R. (2015). Fast R-CNN. In *Proceedings of the IEEE International Conference on Computer Vision*, 1440–1448.

[38] Ren, S., He, K., Girshick, R. and Sun, J. (2015). Faster R-CNN: Towards real-time object detection with region proposal networks. *Advances in Neural Information Processing Systems*, Vol. 28.

[39] Rothe, R., Guillaumin, M. and Van Gool, L. (2015). Non-maximum suppression for object detection by passing messages between windows. In *Computer Vision—ACCV 2014, Lecture Notes in Computer Science*; *2015;* Vol. 9003; pp. 290–306, eds D. Cremers, I. Reid, H. Saito and M.H. Yang. Springer, Cham.

[40] Tabassum, N. and El-Sharkawy, M. (2024). Vehicle detection in adverse weather: A multi-head attention approach with multimodal fusion. *Journal of Low Power Electronics and Applications*, *14*(2), 23.

[41] Redmon, J., Divvala, S., Girshick, R. and Farhadi, A. (2016). You only look once: Unified, real-time object detection. In *Proceedings of the IEEE Conference on Computer Vision and Pattern Recognition*, 779–788.

[42] Redmon, J. and Farhadi, A. (2017). YOLO9000: Better, faster, stronger. In *Proceedings of the IEEE Conference on Computer Vision and Pattern Recognition*, 7263–7271.

[43] Redmon, J. and Farhadi, A. (2018). YOLOV3: An incremental improvement. arXiv preprint arXiv:1804.02767.

[44] Li, L. and Wei, C. (2025). PolarGFusion3D: Polar graph fusion network for enhanced multimodal 3D perception in intelligent vehicles. *IEEE Transactions on Intelligent Vehicles*, *10*, 36–47.

[45] Estrela, V.V., Deshpande, A., Lopes, R.T., Yin, S., Andreopoulos, N., Terziev, A. and Laghari, A. A. (2023). On DICOM, HEVC and 3D medical image compression for volumetric theragnostics. In *Intelligent Healthcare Systems*, ed. V. V. Estrela, 203–216. CRC Press..

[46] Li, L., Wei, C., Sui, S., Hu, J. and Feng, F. (2025). 3-D multiobject tracking with radar data enhancement for intelligent vehicles. *IEEE Sensors Journal*, *25*, 16310–16320.

[47] Estrela, V.V., Aroma, J., Sroufer, R. and Raimond, K., Intorne, A.C., Deshpande, A., Laghari, A.A. and Oliveira, L.P. (2023). Remote sensing applications in disease mapping and public health analysis. In *Intelligent Healthcare Systems*, ed. V.V. Estrela, 185–202. CRC Press.

7 Quantum Computing and Communication for Intelligent Transportation System Networks: Exploring Qubit-based Innovations in Cryptography and Optimization Frontiers

Supriya B.[1*], Srividya Bhat[2], Dhanashree Kulkarni[3], Anand Deshpande[4], Vania V. Estrela[5] and Gabriel Gomes de Oliveira[6]

This chapter examines the growing intersection of quantum computing and Intelligent Transportation Systems (ITS), focusing on secure quantum communication and cryptography that will impact traffic optimization, supply chain, and autonomous vehicles. Quantum technologies have the potential to revolutionize computation and secure communications by leveraging the fundamental principles of quantum mechanics. This text provides a comprehensive overview of quantum computing, covering core concepts such as quantum logic gates, entanglement, and the intrinsic parallelism of qubits. Additionally, the study explores quantum communication protocols, with a particular emphasis on Quantum Key Distribution (QKD) as a mechanism for enhancing data security beyond classical cryptographic techniques. Furthermore, the role of evolutionary algorithms in optimizing quantum systems is examined, including their applications in noise mitigation, circuit optimization, and improving overall system performance. The chapter discusses current challenges in quantum technology, such as decoherence, scalability, and hardware limitations. It outlines future research directions to reach robust, scalable quantum computing and communication infrastructures.

1. Introduction

The rapid development of quantum computing (QC) technology has opened new avenues for addressing complex challenges across many domains of the Intelligent Transportation Systems (ITS), leveraging multiple technologies and applications to boost efficiency, safety, and sustainability in networks spanning all transportation echelons. Traditional computational methods struggle with ITS's

[1,2] Department of Computer Science and Engineering, Sahyadri College of Engineering and Management, Mangaluru, Karnataka, India.
[3] Department of Computer Science & Engineering, Angadi Institute of Technology and Management, (Affiliated to Visvesvaraya Technological University (VTU)), Belagavi, Karnataka, India.
[4] Department of Artificial Intelligence and Data Science, Angadi Institute of Technology and Management, (Affiliated to Visvesvaraya Technological University (VTU)), Belagavi, India.
[5] Telecommunications Department, Fluminense Federal University, Rio de Janeiro, Brazil.
[6] School of Electrical and Computer Engineering, State University of Campinas, Campinas, Brazil.
* Corresponding author: supriyabelagaje@gmail.com

DOI: 10.1201/9781003648147-07

enormous data and complex optimization difficulties. QC can manage and optimize transportation systems through parallel processing, optimization, and machine learning (ML), which has sparked research into bringing QC to ITS. Traffic flow optimization, intelligent course-plotting, autonomous driving, and traffic prediction are promising fields for quantum algorithms (QAs). Quantum concepts (akin to superposition and entanglement) allow these procedures to outperform classical ones in speed, efficiency, and accuracy. Quantum machine learning (QML) may reveal hidden patterns and correlations in large traffic datasets, improving decision-making and prediction. Integrating QC with ITS is difficult since the resulting complexity makes scalable, fault-tolerant quantum device development strenuous. Interdisciplinary teamwork is expected to build QAs for ITS applications, which entail QC and transportation engineering skills. QC's potential benefits for ITS are massive, making integration worth the effort despite these hurdles.

Quantum technologies (QTs) will transform computation and quantum-secure communication in accordance with the laws of quantum mechanics. QC is leading this new revolution by harnessing quantum bits (qubits) that can exist in multiple states simultaneously through superposition [1, 2].

Quantum communication (QComm) protocols are sets of rules and procedures resting on superposition and entanglement for secure and efficient information exchange. They may aid in creating secure cryptographic keys via Quantum Key Distribution (QKD) for classical encryption, or for transmitting unknown quantum states through strategies like quantum teleportation [3, 4].

Metaheuristic algorithms (MAs) optimize depth and gate error of quantum circuits. If a genetic algorithm (GA) rules the mutations and crossovers mimicking biological evolution [5, 6], then a common fitness function might be:

$$Fitness = w_1 \cdot Gate\ Fidelity + w_2 \cdot Circuit\ Depth^{-1}.$$

Here, w_1 and w_2 are weighting coefficients. w_1 controls the importance of achieving high fidelity relative to low circuit depth. A higher w_1 prioritizes accuracy. *Gate Fidelity* maximizes the fidelity to improve the accuracy of the circuit, ensuring it correctly approximates a target quantum operation. w_2 controls the importance of minimizing circuit depth. A higher w_2 prioritizes a shorter, more efficient circuit. *Circuit Depth* minimizes the circuit depth, thus increasing its efficiency, which is crucial for running on error-prone, near-term quantum hardware (QHW).

Besides computation, QComm provides unprecedented advances in data protection. Most notably, QKD forces quantum principles to enable secure key exchange, with any attempted eavesdropping detectable in real time [7]. Eavesdropping causes detectable changes due to quantum collapse during measurement [8, 9]. Hence, the added level of security provided by QComm, compared to traditional cryptographic techniques, makes it a vital resource in the age of cyber-attacks [10, 11, 12]. This promising research has several challenges, including decoherence (quantum information loss), hardware (HW) instability, and scaling QSs stands for Quantum Systems. [5, 13]. To address these, scientists are using MAs in QS design, taking advantage of their flexibility to optimize quantum circuits, reduce noise, and increase reliability [5, 14, 15]. This chapter delves into fundamental concepts for optimizing QComm in ITS and presents the challenges and paths to practical, scalable QC systems.

Quantum computing differs fundamentally from traditional computing (TC) in the storage and processing of information. Rather than mere bits that are either 0 or 1, QC employs qubits that exist in a 0-or-1 state at the same time [1, 16]. This enables quantum computers to solve complex problems in parallel, resulting in incredible computational power for certain tasks. When large numbers of qubits are entangled, they can store and process intricate data structures that increase exponentially with the number of entangled qubits [3].

Quantum logic gates are the QC's key component, which act on qubits much similarly to how classical gates act on bits. Gates include the Hadamard gate (for superposition generation), the Pauli gates (X, Y, Z), and the CNOT gate (for entanglement), among others. Combinations of gates form quantum circuits, which are building blocks of QAs. Efficient circuit synthesis is a major QC problem, mainly due to quantum-state sensitivity [13, 14]. Measurement is a paramount QS feature, in which a qubit's quantum state collapses to a definite classical state (0 or 1). This operation, although necessary for retrieving results from quantum computations, is troublesome because it disturbs the system state and can ruin superposition and entanglement. Additionally, quantum systems are prone to decoherence, the phenomenon in which qubits lose their quantum nature through interactions with the environment [3]. Decoherence imposes a time constraint on computation and is one of the main challenges in developing stable, large-scale quantum computers.

To tackle these issues, researchers are developing Quantum Error Correction (QEC) techniques and investigating HW hinging on superconducting circuits, trapped ions, and topological qubits [7, 8, 11]. Improved QHW and algorithms, along with an understanding of components such as qubits, quantum gates, circuits, measurement, and decoherence, form the foundation for designing both theoretical ideas and practical devices in QC.

Optimization is the central concept of QC, especially in the Noisy Intermediate-Scale Quantum (NISQ) era, when HW limitations and noise become pronounced issues. Perhaps one of the most promising methods to maximize the efficiency and performance of QSs is the use of MAs [17] and ML-reliant optimization models [18].

Ur Rehman et al. [5] present an extensive review of the integration of MAs—GAs [19, 20], Particle Swarm Optimization (PSO) [21, 22], and Differential Evolution (DE) [23, 24]—with QC. These algorithms simulate natural evolution principles, such as selection, crossover, and mutation, to search large solution spaces. In QC, MAs are employed to optimize Quantum Circuit Design (QCD), qubit mapping, gate scheduling, and error mitigation techniques. Their flexibility and adaptability make them especially suitable for NP stands for Nondeterministic Polynomial (time)-hard problems, where classical heuristics fail to scale effectively.

One of the main benefits of employing MAs in QCD is that they can minimize gate depth and error accumulation, both of which are important in the presence of decoherence and HW noise. Evolved circuits can be more faithful while minimizing the number of qubits and QGs stands for Quantum Gates needed, thereby enhancing practical feasibility in current HW architectures.

Following this, Ruiz et al. [13] present AlphaTensor, a model that uses reinforcement learning and deep neural networks to learn optimal matrix-multiplication techniques—a core operation across quantum and classical algorithms. Here, learning-based models can outperform human-designed solutions, and AlphaTensor-type systems can guide automated optimization of quantum circuits, enhancing overall execution efficiency and flexibility.

Vasani et al. [2] emphasize the practical significance of these optimization techniques by presenting their use in QComm networks, cryptographic schemes, and logistics and supply chain management. They provide a comprehensive overview of existing optimization problems and underline that, when integrated with classical MA and learning methods, quantum-enhanced algorithms can hasten progress across various fields.

Collectively, these works illustrate a clear pattern: the merging of metaheuristics and ML with QSs is not purely a theoretical endeavor—it has practical applications in optimizing algorithms, reducing computational costs, and expanding the possibilities of QSs. As QHW develops, these hybrid methods will be essential for realizing scalable, fault-tolerant, and efficient QC and QComm.

2. Recent Contributions on QC and QComm

2.1 Foundations of QC and QComm

Quantum computing is a revolutionary technology that leverages quantum-mechanical principles to perform calculations several orders of magnitude faster than conventional computers. The key component of QC is the qubit, which is the quantum equivalent of the classical bit [1]. Yang et al. [1] gave a complete summary of these fundamental concepts, including control and behavior of qubits, mathematical formalisms for describing quantum operations, and measurement protocols for projecting quantum states onto classical outcomes. Quantum mechanics supports nearly unbreakable data transmission through QKD and other entanglement-based protocols. Architectures for scalable quantum communication systems, such as quantum repeaters, photonic networks, and entanglement-swapping mechanisms, emerge in [4]. The study considers the need for durable quantum infrastructure to enable long-distance communication and identifies current challenges like photon loss, synchronization, and HW constraints. Researchers are working on a future quantum internet to offer ultra-secure, high-speed information exchange globally, with direct applications in cybersecurity, cloud computing, and global communications infrastructure [4]. They not only reveal the theoretical implications of QT but also pave the way for practical applications in cryptography, optimization, and AI.

2.2 Quantum Cryptography and Post-Quantum Security

Quantum machine learning combines QC and artificial intelligence (AI) to handle high-dimensional data and recognize intricate patterns more effectively. It is being used gradually in applications such as image recognition, natural language processing, and bioinformatics [5, 10, 25]. Cryptography remains paramount to both classical and quantum cyber-physical systems [26, 27, 28, 29]. The ongoing work in quantum cryptography is one of the most radical and necessary applications of quantum mechanics. Prominent in this realm is QKD, which allows two parties to generate a shared secret key through the principles of superposition and entanglement, in which attempting to eavesdrop induces detectable disturbances—thus providing built-in security against interception [4, 10].

Quantum computing collaborating with AI in secure communication systems [30] suggests that quantum-enhanced learning models can optimize cryptographic decision-making and threat detection, thereby enabling an intelligent, self-adaptive security framework. Mavroeidis et al. [3] provided an exhaustive analysis of these vulnerabilities and emphasized the urgency of transitioning to Post-Quantum Cryptography (PQC)—algorithms that remain secure against quantum attacks. Easttom [11] advanced lattice-based, multivariate, and hash-based cryptographic systems that hold promise against quantum attacks and are being standardized by bodies such as NIST stands for National Institute of Standards and Technology.

Amanzholova and Priyanka [10] provided an overview of next-generation cryptographic mechanisms that combine quantum-safe protocols with advanced encryption techniques. Their work outlines the key challenges concerning protocol design, implementation, and real-time deployment of quantum cryptographic systems.

Research is actively being driven towards Quantum Digital Signatures, Quantum Oblivious Transfer, and Quantum Secure Direct Communication. These newly proposed protocols are set to redefine secure multi-party communication and will become indispensable in defense, healthcare, and finance, and can involve blockchain-like mechanisms [25, 31, 32, 33, 34].

Lately, with the advent of the post-quantum world, research has continued to focus on hybrid cryptographic infrastructures that amalgamate QHW with classical frameworks for scalable, secure deployment. The integration of quantum-safe protocols, QKD networks, and ML-impelled adaptive encryption represents a major step forward in the cybersecurity landscape [9, 25, 14].

2.3 Quantum Circuit Design (QCD) and Implementations

Quantum Circuit Design and construction lie at the heart of achieving quantum devices' computational potential. Karuppasamy et al. [15] gave an in-depth survey on contemporary approaches for QCD optimization with emphasis placed on three key areas, namely, gate decomposition, noise mitigation strategies, and HW constraints. Ruiz et al. [13] presented AlphaTensor, a deep reinforcement learning architecture that learns new tensor decomposition techniques. These techniques, such as matrix multiplication schemes, surpass classical algorithms originally conceived for classical tensor operations, and carry over directly to quantum circuit compilation. Solar [14] takes this into the cryptographic arena, examining how QCDs affect the performance and efficiency of QAs employed in encryption, hashing, and digital signatures, demonstrating how the layout and organization of circuits directly affect security metrics like nonlinearity, avalanche effect, and key sensitivity, particularly when QSs are utilized to generate or analyze Boolean functions in cryptographic schemes. Together, these studies gesture toward an ever-more data-centric and hybrid paradigm for QCD—where classical optimization algorithms, ML models, and heuristics tuned for quantum circuits blend to provide high-performance, HW-aware designs.

2.4 Quantum Cloud Computing (QCC) and Scalable Architectures

An extensive overview of current Quantum Cloud Platforms (QCPs) from leading vendors, viz., IBM, Google, Amazon Braket, and Microsoft Azure Quantum, appears in [8]. These platforms enable users to execute quantum circuits on real QHW or simulate them on high-performance classical computers. The trend of hybrid quantum-classical computation from [20] involves QCDs performing subroutines of larger classical programs. This approach is most applicable in areas, e.g., optimization, ML, and cryptography, where CSs stands for Classical Systems handle data preprocessing, and QSs execute certain quantum-enhanced operations. In total, Quantum Cloud Computing (QCC) is an essential stepping stone toward large-scale quantum adoption, enabling researchers to prototype, test, and deploy QAs at scale. As HW continues to improve and cloud interfaces become more standardized, this model will be the foundation for future Quantum-as-a-Service (QaaS) ecosystems, making quantum capabilities more accessible and usable for industry and academia [8, 35].

2.5 Qubits and Hardware Innovations

Chae et al. [16] investigated various physical realizations of qubits, discussing how they vary in coherence time, gate velocity, fabrication complexity, and scalability. They also point out how different architectures are adapted to applications like quantum networking, sensing, and large-scale computation. Concerning performance benchmarking, Hu et al. [36] offer real-world evidence of HW capability on D-Wave's 2000-qubit quantum annealer. Empirical evidence for these results shows the operational readiness of mid-scale QCDs for actual work and further emphasizes the necessity of HW-specific algorithm design. Together, these advancements reflect the combined impact of theoretical richness and experimental advances in qubit research. With continued study aimed at improving coherence, reducing crosstalk, and innovating in QEC codes, the science takes its first steps toward realizing scalable quantum architectures capable of running useful workloads in ITS applications.

Physical qubit technologies are the foundation for QHW, each implementation bringing a specific combination of coherence time, scalability, error tolerance, and engineering feasibility. A comprehensive taxonomy of such technologies appears in [16], categorizing them by physical principles, operating environments, and technology readiness. Each of these qubit types comes with trade-offs. Superconducting qubits are easier to scale but require advanced cryogenic infrastructure, whereas trapped ions are more stable but have slower operation times. As Chae et al. [16] point out, there is no "one-size-fits-all" approach available yet. Today's research is highly interdisciplinary and multidisciplinary, drawing on material science, cryogenics, optics, and control theory to extend the limits of each qubit platform. Improvements in hybrid QSs—where several QTs are combined into a single system—are also being investigated to leverage the best features of each platform.

3. Analysis of Quantum Protocols in QComm

This section analyzes and compares popular QComm protocols based on their features, implementation complexity, and challenges. QComm employs fundamental concepts of quantum mechanics to facilitate secure and effective data transmission. Unlike classical communication networks that use mathematical codes for encryption, QComm provides implicit security due to the quantum nature of particles [4, 10]. Since information is in qubits, QComm protocols can identify and prevent eavesdropping, making them handy tools for cybersecurity [10, 12]. QKD, quantum teleportation, and entanglement-based communication are among the most prevalent QComm methods [4, 9].

3.1 QKD Protocols

Quantum Key Distribution is the most advanced and most extensively studied QComm protocol. QKD allows two users to create and exchange a secret cryptographic key, ensuring that any eavesdropping attempt by an adversary to intercept and read it will be detectable. BB84 and E91 QKD protocols are founded on superposition and entanglement principles [4, 10]. Quantum teleportation is another fantastic protocol that enables a quantum state to be transferred from one point to another without transferring the particle [10]. This is done using entangled qubits, shared by both the sender and the receiver. Using a series of quantum operations and classical communication, the quantum state of a particle is rebuilt on a particle at a remote location, in effect "teleporting" the state. Teleportation does not permit faster-than-light communication, but it empowers quantum networking and distributed QC [11]. Entanglement-based communication utilizes the nonlocal correlations of entangled particles to provide secure, synchronized communication channels. If two parties share an entangled pair of qubits, then what happens on one qubit instantly influences the other, irrespective of the distance between them. This property is utilized in protocols such as entanglement swapping and quantum repeaters, which are the basis for constructing long-distance quantum networks and quantum internet infrastructure [4, 25]. All of these quantum communication protocols represent a quantum leap over traditional cryptographic techniques, promising enhanced privacy, greater security against cyberattacks, and the potential to develop highly secure global communications networks.

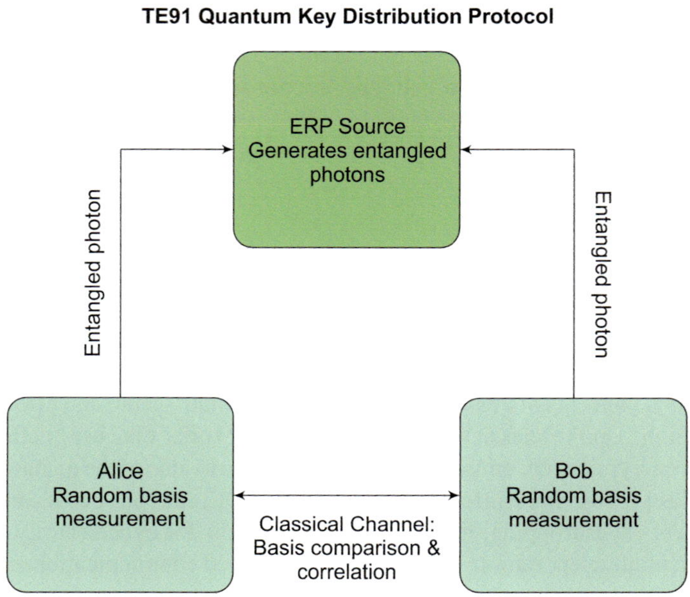

Fig. 7.1(a) TE91 Quantum Key Distribution protocol

Quantum Key Distribution stands out as one of the most developed and practical uses of QComm. It allows two parties to create a shared, secret cryptographic key over a potentially insecure quantum channel, with its security backed by the quantum principles. Unlike traditional encryption methods that depend on complex mathematics, QKD protocols provide unconditional security—any eavesdropping attempt causes noticeable disturbances in the quantum states being sent. Among them, the most recognized QKD protocols are BB84 and E91. E91 Protocol, which was formulated by Ekert in 1991, employs entangled pairs of qubits rather than the transmission of single qubits. Alice and Bob are given half of an entangled pair each and measure them in randomly selected bases, as shown in Figure 7.1(a). Correlations between their outcomes, verified using Bell-test inequalities, are used to produce a common key. The approach expands security grounded on quantum entanglement and is automatically immune to certain attacks [4]. The BB84 Protocol, introduced by Bennett and Brassard in 1984, relies on the transmission of qubits within randomly selected non-orthogonal bases (e.g., rectilinear and diagonal). The sender (Alice) encodes qubits with one of these bases, and the receiver (Bob) measures them with randomly selected bases, as depicted in Figure

7.1(b). Once transmitted, they publicly match the bases (not the values) used and retain the bits that employ the same basis. Any eavesdropping attempt introduces detectable errors due to the no-cloning theorem and the collapse of the quantum state during measurement [9].

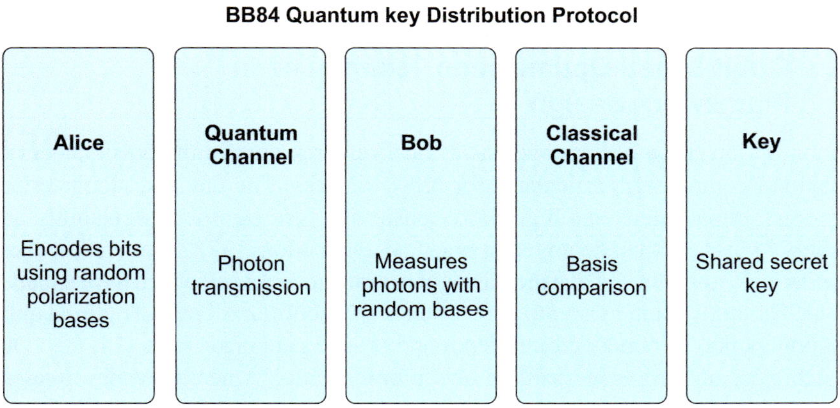

Fig. 7.1(b) BB84 Quantum Key Distribution protocol

Hasan et al. [4] examine both protocols under real-world implementation scenarios, including photon loss, channel noise, and detector inefficiencies. They highlight the imperative of quantum repeaters and entanglement purification methods in taking QKD to long-distance networks. Based on the quantum, QKD should be a crucial part of creating quantum-secured networks that underpin emerging applications, e.g., the quantum internet, government-grade encryption, and secure space communication. As QC grows, its incorporation into cybersecurity architectures is transforming the world of cryptographic models and information assurance. Alabi [7] describes the paradigm shift initiated by QC in cryptographic theory and real-world applications. His work highlights that quantum cryptography is not merely a defensive technology but also a revolutionary technology that enables new modes of secure communication—e.g., quantum-secure blockchain, zero-knowledge proofs, and quantum-boosted authentication mechanisms.

Golec et al. [8] build on this by emphasizing the contribution of quantum cryptographic technologies to cybersecurity policy and business planning. Their work outlines how quantum cryptography could be strategically implemented across industries to support data protection, secure cross-border communication, and future-proof against emerging threats. From an algorithmic and HW perspective, Hu et al. [36] present the first application of quantum annealing on D-Wave's 2000-qubit chip for producing cryptographic Boolean functions. These are essential in stream and block ciphers, digital signatures, and hashing algorithms because they demonstrate how quantum annealing can crack combinatorial optimization problems in cryptography. Together, these advances represent a trend toward quantum-integrated cryptography, where both defensive (QKD, quantum random number generators) and generative (e.g., quantum function synthesis, cryptographic HW design) methods are being investigated. This represents a significant shift from protecting against quantum attacks to actively utilizing quantum mechanics in designing next-generation cryptographic ecosystems.

4. Qubit-based Optimization Techniques in Framework Design

Qubit optimization has emerged as a focal research area in the development of quantum computing, particularly for NISQ devices. The devices, characterized by small qubit counts and high noise sensitivity, face significant scalability and reliability issues when deployed in practical applications [37]. To overcome these limitations, researchers have increasingly resorted to MAs [5], which have excelled at QCD optimization by identifying efficient QG configurations and optimal qubit mappings that minimize circuit depth and reduce gate error rates [14, 15]. One notable use of MAs is in quantum circuit optimization, where they aim to lessen circuit depth, gate count, and decoherence vulnerability. Ruiz et al. [13] showed that MA-based models like AlphaTensor can optimize tensor computations in quantum circuits, yielding efficient matrix decompositions with fewer operations, which directly enhances QA performance and accuracy [13]. Such strategies enable the evolution of quantum SW stands for Software that better fits the limitations of existing NISQ HW.

In addition, not only does MAs integration with QC deal with optimization challenges, but it also creates novel paradigms in hybrid classical-quantum computing. Ur Rehman et al. [5] explain how hybrid paradigms fuse the global searching ability of MAs with quantum subroutines to solve immensely complex, NP-hard challenges [2].

As QHW advances, particularly with the emergence of quantum cloud services and shared computing infrastructure [8], the necessity of intelligent optimization methods such as MAs cannot be overstated. MAs present an adaptable, robust, and scalable optimization paradigm that supports the distinctive requirements of QC. Their very flexibility enables scientists to optimize QCD performance, suppress noise, calibrate HW, and even create novel quantum protocols. The compatibility of MAs and QC thus offers a promising avenue toward transcending today's technological limits and achieving the full potential of QTs during the NISQ era and beyond [5, 6, 13, 14, 15].

4.1 Challenges and Limitations

Quantum computing and QComm technologies hold great promise, but their development and deployment face significant challenges. Overcoming these challenges requires concerted efforts in QCD, algorithm development, and QS optimization. Advances continue to push the limits, making scalable, fault-tolerant QSs a reality.

- *Decoherence and Noise*: Quantum systems are susceptible to the environment. Decoherence—the destruction of quantum coherence stemming from environmental interactions—is still one of the most significant caveats in creating stable quantum systems [1, 3]. Noise degrades gate fidelity and imposes an upper bound on the time available for useful QC.
- *Qubit Scalability and Connectivity*: Scaling QSs to accommodate thousands or millions of qubits calls for innovation in hardware, interconnects, and controls. The majority of existing architectures are limited in how qubits communicate with one another, which prevents the implementation of algorithms at scale [7, 16].
- *Error Correction Overhead*: While QEC offers resilience, it imposes significant overhead. One logical qubit could need hundreds to thousands of physical qubits for fault tolerance, dramatically increasing HW demands [14, 13].
- *QHW Fragility*: Physical systems upon which qubits are implemented, e.g., superconducting circuits or trapped ions, are sensitive and must have strict operational requirements such as cryogenic temperatures or ultra-high vacuum conditions [8].
- *Algorithm and SW Limitations*: Even with advances in theory, few QAs today have a definite advantage over their classical equivalents. General-purpose, noise-tolerant QAs are still a primary research focus [5, 10].
- *Security Concerns in Early Quantum Networks*: Although QComm is said to provide secure data exchange, initial deployments are exposed to threats such as side-channel attacks, device imperfections, and realistic eavesdropping attacks [4, 9]. Secure implementation protocols and standards are still being developed.

4.2 ITS Applications and Use Cases Using QC, QComm, and Quantum Cryptography

Quantum technologies are gaining traction across industries, with strong applications and creative use cases, highlighting the revolutionary potential of QC and QComm systems to address complex, real-world challenges. QTs can enhance ITS by enabling secure communication via quantum cryptography, thereby protecting sensitive data in transit. QC supports advanced computation, allowing the solution to complex transportation problems in real time. QComm improves sensing, actuation, and networking, enabling accurate data sharing between cyber-physical systems and digital twin layers.

- *Drug Discovery/Delivery and Material Science*: QC's capacity to precisely simulate molecular configurations and chemical processes has opened up new horizons in drug and materials development [38, 39, 40]. By accurately modeling intricate molecules, QSs can speed up the discovery of drug candidates and new compounds [8, 15]. QTs can also revolutionize personalized drug delivery and theragnostics [41, 42, 43].

- *Financial Modeling and Risk Analysis*: QAs provide enhanced modeling and optimization functions for ITS financial markets. Applications involve portfolio optimization, risk analysis, and fraud detection by considering multiple scenarios simultaneously [5].
- *Logistics and Supply Chain Optimization*: Logistics problems (including vehicle routing, inventory management, and scheduling [44, 45, 46, 47]) are amenable to quantum optimization methods, such as the Quantum Approximate Optimization Algorithm (QAOA). These solutions can dramatically cut costs and improve operational efficiency [8, 13]. QC systems available in the cloud, like IBM Quantum, Amazon Braket, and Google Cirq, allow international collaboration, experimentation, and education. These systems offer simulated environments and access to real HW for researchers, developers, and students [1, 8], capturing the extensive QT potential in addressing computationally intensive, highly secure, and analytically challenging problems. With ongoing improvements in HW and algorithms, quantum applications will expand into even more domains.

4.3 Future Directions

The QC and communication directions focus on realizing scalable, fault-tolerant, and practically useful systems. Perhaps the most challenging aspiration is the implementation of a solid, functional quantum internet for ultra-secure communication with entanglement-based protocols and quantum repeaters, changing worldwide data exchange [2, 48]. Another urgent direction is the development of fault-tolerant QC. Modern QSs are enormously noisy and decoherent. But with QEC and logical qubits, one can correct these problems and execute trustworthy long-term calculations [13]. QML will also pick up momentum. It combines QC with AI to analyze massive datasets, identify patterns, and enable more effective models in applications such as healthcare, cybersecurity, opinion mining, and climate science [14, 49]. Hardware also promises to make headways in superconducting qubits, trapped ions, and topological qubits, each with its specific strengths in scalability, coherence time, and operational speed [8, 16]. The use of MAs remains a promising direction for automating circuit optimization, minimizing resource overhead, and dynamically adjusting hardware configurations [5]. Cloud-based quantum platforms and hybrid classical-quantum frameworks will also be key to mainstream adoption by making resources accessible and filling performance gaps [1, 8]. With continued interdisciplinary research, public and private sector funding, and global collaborations, quantum technologies are poised to revolutionize cryptography, logistics, drug discovery, and more [3, 7].

5. Conclusion

Quantum technologies are revolutionizing the future of ITS by revolutionizing computing and secure communication [50, 51, 52, 53, 54, 55]. By applying the basic principles of superposition, entanglement, and quantum coherence, QC has the potential to surpass classical computing in cryptography, optimization, simulation, and AI, among other realms [1, 7]. Similarly, QComm protocols, particularly QKD, promise unmatched security grounded in the laws of physics rather than computational assumptions [4, 10]. Despite extensive advances in technology, several challenges remain: decoherence and error correction overhead in hardware scalability and algorithm design [16, 48]. Nonetheless, the application of MAs and the growing availability of quantum cloud platforms are rapidly closing the gap between theory and practical application [5, 8]. Future research and development are expected to push boundaries even further, aiming to create a quantum internet, fault-tolerant QC, and efficient QML models [2, 48]. As interdisciplinary research continues to converge and industries start implementing quantum solutions, the vision of accessible, scalable, and application-ready quantum systems becomes increasingly real. Finally, the development of QTs offers transformative possibilities across fields, envisioning a new paradigm for how problems are solved, data are secured, and knowledge is amplified [3].

Acknowledgments

We extend our deepest gratitude to the instructors of the Department of Computer Science and Engineering at Sahyadri College of Engineering and Management, Mangaluru, for their constructive feedback and motivation in preparing this chapter. We thank the reviewers for their insightful comments, which helped enhance the quality of this work. We also recognize the authors and researchers of the works cited, as well as the open-source cryptography and communication research communities, whose original work has heavily influenced and inspired this research on quantum technologies.

Availability Statement

There were no new data generated, analyzed, or reported in this chapter. The discussion is based solely on published literature, which has been appropriately cited throughout the manuscript. All cited articles are publicly accessible and included in the references for future perusal.

References

[1] Yang, Z., Zolanvari, M. and Jain, R. (2023). A survey of important issues in quantum computing and communications. *IEEE Communications Surveys & Tutorials 25*(2), 1059–1094.

[2] Vasani, V., Prateek, K., Amin, R., Maity, S. and Dwivedi, A.D. (2024). Embracing the quantum frontier: Investigating quantum communication, cryptography, applications and future directions. *Journal of Industrial Information Integration*, *39*, 100594.

[3] Mavroeidis, V., Vishi, M., Zych, M. and Jøsang, A. (2018). The impact of quantum computing on present cryptography. *arXiv preprint* arXiv:1804.00200.

[4] Hasan, S.R. et al.(2023). Quantum communication systems: Vision, protocols, applications, and challenges. *IEEE Access*, *11*, 15855–15877.

[5] ur Rehman, J. et al. (2025). Evolutionary algorithms and quantum computing: Recent advances, opportunities, and challenges. *IEEE Access*.

[6] Krawec, W., Picek, S. and Jakobovic, D. (2019). Evolutionary algorithms for the design of quantum protocols. In *Applications of Evolutionary Computation*, eds P. Kauffman & P. Castillo, 233–248. Springer.

[7] Alabi, M. (2024). *Quantum Computing: A Paradigm Shift in Computational Power and its Potential Applications*. ResearchGate.

[8] Golec, M., Hatay, E.S., Golec, M., Uyar, M., Golec, M. and Gill, S.S. (2024). Quantum cloud computing: Trends and challenges. *Journal of Economy and Technology*, *2*, 190–199.

[9] Minhas, N.N. (2024). A survey on quantum cryptography, its protocols, applications, and challenges. *Preprints*, 2024080333.

[10] Amanzholova, S. and Priyanka, A.C. (2025). Exploring advancements, applications, and challenges in the realm of quantum cryptography. In *Next Generation Mechanisms for Data Encryption*, 116.

[11] Easttom, C. (2022). Quantum computing and cryptography. In *Modern Cryptography: Applied Mathematics for Encryption and Information Security*, 397–407. Cham: Springer International Publishing.

[12] Sood, N. (2024). Cryptography in post quantum computing era. *Available at SSRN*, SSRN ID: 4705470. Available at https://papers.ssrn.com/abstract=4705470. Accessed on 03 February 2026.

[13] Ruiz, F.J.R., Laakkonen, T., Bausch, J. et al. (2025). Quantum circuit optimization with AlphaTensor. *Nature Machine Intelligence*, *7*, 374–385.

[14] Solar, M. (2024). A review on quantum machine learning and quantum cryptography. *Memoria Investigaciones en Ingenierıa*, *27*, 180–199.

[15] Karuppasamy, K., Puram, V., Johnson, S. and Thomas, J.P. (2025). A comprehensive review of quantum circuit optimization: Current trends and future directions. *Quantum Reports*, *7*(1), 2.

[16] Chae, E., Choi, J. and Kim, J. (2024). An elementary review on basic principles and developments of qubits for quantum computing. *Nano Convergence*, *11*, 11.

[17] Estrela, V.V., Intorne, A.C., Batista, K.K.S., Deshpande, A., Aroma, J., Raimond, K.. Shi, F, Laghari, A.A. and Lin, Y.-D. (2023). Metaheuristics applied to pathology image analysis. In *Intelligent Healthcare Systems*, ed. V. V. Estrela, 322–340. CRC Press.

[18] Urgelles, H., Garcia-Roger, D. and Monserrat, J. F. (2024). Quantum-based maximum likelihood detection in MIMO-NOMA systems for 6G networks. *Quantum Reports*, *6*(4), 533–549.

[19] b. França, R.P., Monteiro, A.C.B., Estrela, V.V. and Razmjooy, N. (2021). Using metaheuristics in discrete-event simulation. In *Metaheuristics and Optimization in Computer and Electrical Engineering. Lecture Notes in Electrical Engineering, vol 696*, eds N. Razmjooy, M., Ashourian and Z. Foroozandeh. Springer, Cham.

[20] Martinez E.G.F. and Zich, R.E. (2024). Novel quantum computation based selection operator for genetic algorithms applied to electromagnetic problems. In *2024 Proceedings of the 18th European Conference on Antennas and Propagation (EuCAP)*, IEEE, Glasgow, UK, 1–4.

[21] Estrela, V.V., Razmjooy, N., Monteiro, A.C.B., França, R.P., de Jesus, M.A. and Iano, Y. (2021) A computational intelligence perspective on multimodal image registration for unmanned aerial vehicles (UAVs). In *Metaheuristics and Optimization in Computer and Electrical Engineering. Lecture Notes in Electrical Engineering, vol 696*, eds N. Razmjooy, M. Ashourian and Z. Foroozandeh. Springer, Cham.

[22] Chaabane, L., Khelassi, A., Terziev, A., Andreopoulos, N., de Jesus, M.A. and Estrela, V.V. (2021). Particle swarm optimization with tabu search algorithm (PSO-TS) applied to multiple sequence alignment problem. In *Advances in Multidisciplinary Medical Technologies—Engineering, Modeling and Findings*, eds A. Khelassi & V.V. Estrela. Springer, Cham.

[23] Dixit, A., Mani, A. and Gorbachev, S. (2024). Hybrid framework for medical image classification: Integrating quantum DE and deep learning. *2024 IEEE International Conference on Intelligent Signal Processing and Effective Communication Technologies (INSPECT)*, 1–6.

[24] Zufan, P. and Bidlo, M. (2021). Advances in evolutionary optimization of quantum operators. *MENDEL*, *27*(2), 12–22.

[25] Kulkarni, D.S., Deshpande, A.B., Bhandurge, S.A., Estrela, V.V. and Deshpande, A. (2025). Impact of textual length on lexicon-based sentiment classification for Hindi reviews. *Cureus Journal of Computer Science*, *2*, 2025.

[26] de Jesus, M.A., Laghari, A.A., Khan, A.A., Jumani, A.K., Shabaz, M., Deshpande, A., Jenice Aroma, R., Raimond, K. and Khan, A. (2024). Security in blockchain-based smart cyber-physical applications relying on wireless sensor and actuators networks. In *Applying Artificial Intelligence in Cybersecurity Analytics and Cyber Threat Detection*, eds S. Mahajan, M. Khurana and V. V. Estrela.

[27] Stutz, D., de Assis, J.T., Laghari, A.A., Khan, A.A., Deshpande, A., Kulkarni, D., Terziev, A., de Jesus, M.A. and Grata, E.G.H. (2024). Cyber threat detection and mitigation using artificial intelligence—A cyber-physical perspective. In *Applying Artificial Intelligence in Cybersecurity Analytics and Cyber Threat Detection*, S. Mahajan, M. Khurana and V. V. Estrela.

[28] Laktionov, I., Diachenko, G., Moroz, D. and Getman, I. (2025). A comprehensive review of cybersecurity threats to wireless infocommunications in the quantum-age cryptography. *IoT*, *6*(4), 61.

[29] NIST. Status Report on the Fourth Round of the NIST Post-Quantum Cryptography Standardization Process. Available at https://csrc.nist.gov/pubs/ir/8545/final (accessed on 11 March 2025).

[30] Ahmadi, A. (2023). Quantum computing and artificial intelligence: The synergy of two revolutionary technologies. *Asian Journal of Electrical Sciences*, *12*(2), 15–27.

[31] Khan, A.A., Laghari, A.A., Almansour, H. et al. (2025). Quantum computing empowering blockchain technology with post quantum resistant cryptography for multimedia data privacy preservation in cloud-enabled public auditing platforms. *Journal of Cloud Computing*, *14*, 43.

[32] Estrela, V.V., de Jesus, M.A., Intorne, A.C., Batista, K.K.S., Deshpande, A., Shi, F., Khan, A.A. and Oliveira, L.P. (2023). Blockchain technology enabling better services in the healthcare domain. In *Intelligent Healthcare Systems*, ed. V.V. Estrela, 135–158. CRC Press.

[33] Bukhsh, K., Rind, M.M., Shaikh, S., Khan, A.A., Laghari, A.A., Estrela, V.V. and Andreopoulos, N. (2025). Blockchain and Internet of Things (IoT) enabled secure communication protocols for smart agriculture. In *Intelligent Designs, Innovations and Sustainability in Agriculture 4.0*, eds A. Deshpande, V. V. Estrela and A. Khan, 267–284. CRC Press.

[34] Kearney, J.J. and Perez-Delgado, C.A. (2021). Vulnerability of blockchain technologies to quantum attacks. *Array*, *10*, 100065.

[35] Maass, W., Agrawal, A., Ciani, A., Danz, S., Delgadillo, A., Ganser, P., Kienast, P., Kulig, M., König, V., Rodellas-Gracia, N., Rughubar, R., Schröder, S., Stautner, M., Stein, H., Stollenwerk, T., Zeuch, D. and Wilhelm, F.K. (2024). Quantum computing enhanced service ecosystem for simulation in manufacturing. ArXiv, abs/2401.10623.

[36] Hu, F., Lamata, L., Sanz, M., Chen, X., Wang, C. and Solano, E. (2020). Quantum computing cryptography: Finding cryptographic Boolean functions with quantum annealing by a 2000 qubit D-Wave quantum computer. *Physics Letters A*, *384*(10), 126214.

[37] Preskill, J. (2018). Quantum computing in the NISQ era and beyond. *Quantum*, *2*, 79.

[38] Egan, L., Debroy, D.M., Noel, C., Risinger, A., Zhu, D., Biswas, D., Newman, M., Li, M., Brown, K.R., Cetina, M. and Monroe, C.R. (2020). Fault-tolerant operation of a quantum error-correction code. arXiv: Quantum Physics.

[39] Estrela, V.V., Bukhsh, K., Malook Rind, M., Shaikh, S., Khan, A.A., Laghari, A.A., Deshpande, A., Andreopoulos, N. and Terziev, A. (2023). Deep learning as a driving force for better drug development. In *Intelligent Healthcare Systems*, ed. V.V. Estrela, 217–235, CRC Press.

[40] Estrela, V.V., de Jesus, M.A., Intorne, A.C., Batista, K.K.S., Deshpande, A., Shi, F., Laghari, A.A., Khan, A.A. and Oliveira. L.P. (2023). Medical visual theragnostic systems using artificial intelligence (AI)—Principles and perspectives. In *Intelligent Healthcare Systems*, ed. V.V. Estrela, 301–321. CRC Press.
[41] Sutradhar, K., Venkatesh, R. and Venkatesh, P. (2025). A review on smart healthcare employing quantum Internet of Things. *IEEE Engineering Management Review*, *53*, 141–153.
[42] Sangtani, A., Nag, O.K., Oh, E. et al. (2021). Quantum dot-enabled membrane-tethering and enhanced photoactivation of chlorin-e6. *Journal of Nanoparticle Research*, *23*, 159.
[43] Mauranyapin, N.P., Terrasson, A. and Bowen, W.P. (2022). Quantum biotechnology. *Advanced Quantum Technologies*, *5*, 2100139.
[44] Khan, A.A. et al. (2025). D-Agri: A solution for smart farming using AI & drone technology. *2025 IEEE 2nd International Conference on Electronics, Communications and Intelligent Science (ECIS)*, Yueyang, China, 1–6.
[45] Estrela, V.V., Deshpande, A., Stutz, D., de Assis, J.T., Jumani, A.K., Khan, A.A., Shi, F., Yin, S. and Lin, Yu-Da (2023). Ad hoc networks in healthcare intelligent transportation systems humanitary relief—MANETs, VANETs, and FANETs. In *Intelligent Healthcare Systems*, ed. V. V. Estrela, 66–89. CRC Press.
[46] Wurtz, J., Sack, S. and Wang, S. (2024). Solving nonnative combinatorial optimization problems using hybrid quantum–classical algorithms. *IEEE Transactions on Quantum Engineering*, *5*, 1–14.
[47] Wagner, F., Nüßlein, J. and Liers, F. (2023). Enhancing quantum algorithms for quadratic unconstrained binary optimization via integer programming. *ACM Transactions on Quantum Computing*, *6*, 1–28.
[48] Wehner, S., Elkouss, D. and Hanson, R. (2018). Quantum internet: A vision for the road ahead. *Science*, *362*, eaam9288.
[49] Kulkarni, D.S. and Rodd, S.F. (2018). Extensive study of text based methods for opinion mining. In *2018 2nd International Conference on Inventive Systems and Control (ICISC)*, 523–527. IEEE.
[50] Singh, J., Tyagi, N., Singh, S., Alzubi, A.A., AlZubi, F.I., Sehra, S.S. and Ali, F. (2023). Enhancing indoor navigation in intelligent transportation systems with 3D RIF and Quantum GIS. *Sustainability*. Sustainability 2023, volume 15, issue 22, article 15833 (pp. 1–17)
[51] Karuna, G., Ram, R.B., MuhsnHasan, M., Reddy, K.P., Surya, E. and Niranjanamurthy, M. (2025). Enhancing intelligent transport systems through quantum and blockchain security in vehicular networks. *2025 International Conference on Computational Innovations and Engineering Sustainability (ICCIES)*, 1–7.
[52] Pei, J., Wang, L., Awan, N. and Alturki, R. (2025). Advancing federated learning privacy with quantum communication techniques: A robust scalable framework. *IEEE Systems, Man, and Cybernetics Magazine*, *11*, 51–58.
[53] Singh, R., Das, L. and Chaudhary, D. (2025). Building scalable and quantum attack resistant authenticated message delivery system for internet of vehicles with blockchain consensus mechanism. *Concurrency and Computation: Practice and Experience*, *37*(9–11), e70071.
[54] Zhang, Z., Cao, Z. and Wang, Y. (2025). Forensics system for internet of vehicles based on post-quantum blockchain. *Sensors*, *25*(19), 6038.
[55] Zhuang, Y., Azfar, T., Wang, Y., Sun, W., Wang, X., Guo, Q. and Ke, R. (2024). Quantum computing in intelligent transportation systems: A survey. ArXiv, abs/2406.00862.

8 Generative AI-based Framework for Road Infrastructure Design in Intelligent Transportation Systems

Sonam Bhandurge[1*], Kuldeep Sambrekar[2], Priyanka Pujari[3], Dhanshree Kulkarni[4], Asiya Khan[5] and Stephany Berrio Perez[6]

This chapter explores the transformative role of generative artificial intelligence (AI) and intelligent methodologies in modern road infrastructure design for urban environments facing rapid population growth and evolving mobility demands. It critically examines the limitations of traditional, human-centric road design approaches and chronicles the historical evolution of transportation systems, highlighting the progression from reactive strategies to adaptive, technology-driven solutions. The chapter details the integration of Intelligent Transportation Systems (ITS), spotlighting AI, deep learning, and generative models including Generative Adversarial Networks (GANs), Variational Autoencoders (VAEs), and diffusion models as drivers of automated, multi-objective optimization in road network planning. It introduces a comprehensive framework encompassing data collection, preprocessing, generative modeling, simulation-based evaluation, and iterative feedback that collectively enables the synthesis, assessment, and refinement of robust, context-aware infrastructure layouts. The framework demonstrates significant advancements in traffic efficiency, safety, and sustainability. The findings highlight how AI-based generative design and reinforcement learning have the potential to build inclusive, adaptable urban mobility networks, a paradigm shift in city thinking and execution of future-proof road networks.

1. Introduction

Currently, more than half of the world's population is already urbanized, and the number is anticipated to reach nearly 70% by 2050. This sharp rise in urban population poses enormous pressure on the infrastructure, primarily transportation networks. Cities as hubs of economic growth and social interaction are continually confronting problems such as chronic traffic jams, long travel times, increased

[1,2] Department of Computer Science and Engineering, Gogte Institute of Technology, Belagavi, Karnataka, India.
[3,4] Department of Computer Science and Engineering, Angadi Institute of Technology and Management, (Affiliated to Visvesvaraya Technological University (VTU)), Belagavi, Karnataka, India.
[5] University of Plymouth, Plymouth, United Kingdom.
[6] Department of Australian centre of field robotics, University of Sydney, Sydney, Australia.
* Corresponding author: sonambhandurge.aitm@gmail.com

DOI: 10.1201/9781003648147-08

carbon output, and greater accident rates. Not only do these affect the standard of living for citizens, but they also exert enormous economic costs in lost productivity and increased use of resources.

Road design in the past has been a highly iterative, resource-intensive, and traditional process. Conventional approaches normally consist of a step-by-step workflow, with feasibility studies first, conceptual design next, followed by detailed engineering, and then construction. This is a highly dependent approach on human designers' abilities and judgment using a variety of traditional tools such as CAD packages, Geographic Information Systems (GIS), and engineering manuals. Even though such approaches have had their day, they have inherent faults in handling the complexities of today's infrastructure demands.

1.1 Historical Background of Transportation Systems

The evolution of transportation systems has consistently mirrored the advancement of human civilizations. From the early trade pathways and primitive road networks of ancient societies to the emergence of extensive national highways and expressways in the twentieth century, road infrastructure has remained fundamental to socioeconomic development [1]. As highlighted across diverse domains of research—including linguistic technology [2, 3] and secure data management for scalable systems [4]—systematic improvements in foundational structures are crucial for supporting large-scale growth. Similarly, the continual enhancement of transportation frameworks plays a transformative role in enabling connected, resilient, and future-ready communities [1].

Early projects like PROMETHEUS and PATH in the 1980s and the 1990s set the groundwork for the historical evolution of Intelligent Transportation Systems (ITS) [1]. In order to improve road safety and efficiency through infrastructure–vehicle communication, PROMETHEUS used sensors such as radar, laser scanners, and cameras. Around the same period, the US PATH project focused on optimizing highway traffic using roadside sensors and early vehicle platooning tests. Technologies like SAFESPOT, as the 2000s progressed, allowed vehicles and roadside equipment in a series of European cities to exchange data in real time, further enabling early threat detection and traffic awareness. Through the assistance of embedded intelligence and communication technology, these innovations marked the transition from autonomous systems toward collaborative, real-time traffic management.

Human needs and technology advancements have continuously affected the evolution of transportation systems throughout history [5]. It started with the building of fundamental infrastructure, such as trade routes and roadways, which allowed early civilizations to develop, communicate, and engage beyond tribal lines. Vehicles like cars, trains, and ships were invented and widely used as societies developed, revolutionizing mobility and facilitating faster and easier access to both personal and commercial travel.

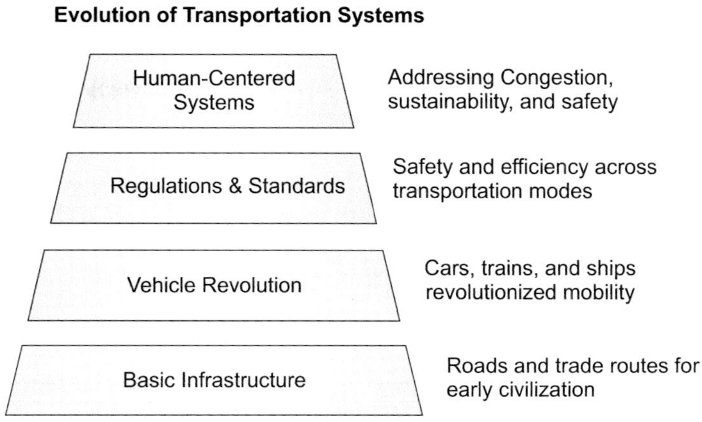

Fig. 8.1 Evolution of transportation systems

As transportation networks became more complicated, rules and guidelines were put in place to guarantee efficiency, safety, and coordination across various modes, including air, sea, rail, and road. The emphasis has evolved in recent decades toward developing more human-centered systems that solve problems including public safety, environmental sustainability, and transportation congestion. In addition to facilitating mobility, modern transportation improves quality of life and is essential in forming economies, cities, and social relationships.

1.2 Transitioning from Conventional Routes to Digital Routes

Traffic was first managed via fixed infrastructure, such as induction loops and traffic lights, which had little real-time flexibility. With the incorporation of smart roadside infrastructure, including cameras, LiDAR, radar, and roadside units (RSUs), which gather real-time traffic data, these systems gradually developed into more sophisticated versions [1]. Today, this information is utilized to build digital twins of transportation environments that allow for virtual monitoring and management of traffic conditions, vehicle dynamics, and safety status.

Traditional physical routes are being replaced by smarter, digital ones in transportation systems. Transportation is becoming more data-driven and dynamic due to technological advancements. These days, real-time data from digital sources such as GPS, smartphone apps, and internet platforms is used to manage flows, direct traffic, and influence traveller behavior [5].

1.3 Need for Intelligent Transportation Systems (ITS)

As cities grow more crowded and traffic problems become more complex, the need for smarter, more adaptive road systems becomes urgent. That is where ITS comes in. Formally, ITS refers to a network of advanced technologies including sensors, communication devices, software, and control systems, designed to make transportation safer, more efficient, and more responsive. ITS is framed as a technology backbone for modern urban mobility solutions. At its core, ITS brings together components like roadside infrastructure, in-vehicle systems, data centers, and cloud or edge computing platforms, all working together to monitor, analyze, and manage traffic in real time. ITS refers to the use of artificial intelligence (AI), computer vision, and remote sensing technologies to optimize urban planning, analyze crowd and traffic behavior, predict traffic conditions, and enhance smart city development.

The main components of ITS include [1]:

- *Sensors and Perception Systems*: ITS relies heavily on a wide array of sensors such as cameras, radar, LiDAR (Light Detection and Ranging), weather sensors, and induction loops embedded in roads. These sensors capture real-time data about traffic conditions, vehicle positions, road environment, and anomalies.
- *Roadside Infrastructure (RSUs)*: RSUs are fixed devices installed on traffic lights, gantries, or street poles. They serve as critical nodes for gathering sensor data and enabling vehicle-to-infrastructure (V2I) and infrastructure-to-vehicle (I2V) communication.
- *Communication Technologies*: ITS uses wireless and network communication protocols such as 5G, Long Term Evolution (LTE), Dedicated Short-Range Communications (DSRC), and fiber optics to allow vehicle-to-everything (V2X) connectivity. This enables vehicles to communicate not only with infrastructure but also with each other and with cloud-based systems.
- *Edge and Cloud Computing Systems*: These components handle real-time data processing either locally (edge computing at RSUs) or centrally (cloud systems), supporting functions like traffic prediction, hazard detection, and dynamic control.
- *Control and Actuation Systems*: Based on processed data, ITS can dynamically adjust traffic lights, activate warning signs, or influence vehicle behaviors to optimize flow and ensure safety.
- *Central Traffic Management Centers*: These centers act as the supervisory backbone, overseeing the coordination between distributed ITS elements. They integrate data, manage response strategies, and deliver insights to both operators and automated systems.
- *User Interaction Interfaces*: ITS also includes systems that interface with end users—such as in-vehicle displays, smartphone applications, or variable message signs (VMS)—to inform, alert, or guide road users in real time.

Together, these components enable ITS to deliver services such as hazard warnings, adaptive routing, traffic light optimization, emergency response coordination, and data-driven urban mobility planning.

1.4 Smart Services Enabled by Intelligent Transportation Systems

In order to actively improve road safety, lessen traffic, and enhance the entire travel experience, modern ITS provide a variety of sophisticated functions that go beyond simple traffic control. One such service is emergency vehicle alerting, which helps traffic systems prioritize and identify the passage of ambulances, police cars, and fire engines to speed up response times and clear the path. Another helpful feature is the Green Light Optimal Speed Advisory (GLOSAR), which suggests the best driving speeds to reduce delays and fuel consumption by allowing automobiles to pass many green lights without stopping. ITS also improves driving safety and situational awareness through features such as obstacle detection and blind spot warning, improving safety against accidents in congested traffic conditions. With increasing acceptance of automation, systems currently enable autonomous parking aid, allowing cars to park themselves safely with minimal to no intervention by the driver. Moreover, real-time route guidance employs current traffic information to suggest the most effective paths of travel, enabling drivers to bypass congested zones, save time, and reduce emissions. These services, as shown in Figure 8.2, illustrate how ITS makes transportation smarter, safer, and more responsive that is customized for both day-to-day drivers and long-range urban mobility objectives.

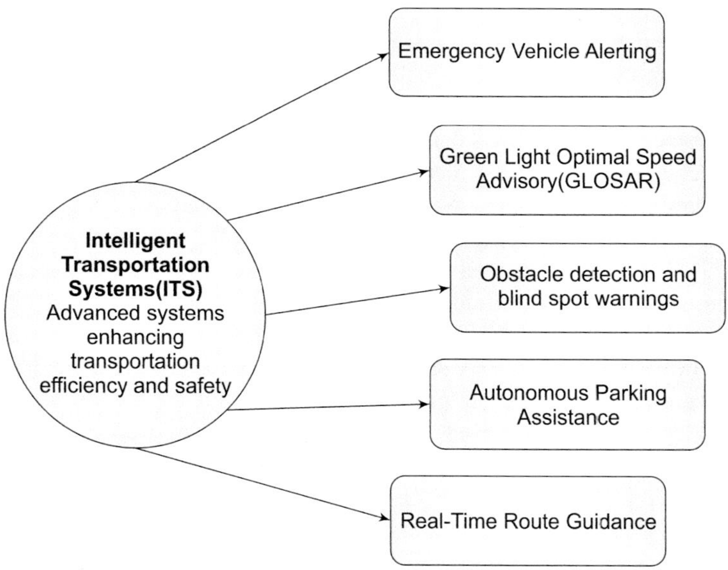

Fig. 8.2 Multifaceted benefits of ITS

1.5 Limitations of Conventional ITS and the Rise of Generative AI

Conventional ITS models struggle significantly in scenarios where data is sparse, uncertain, or highly variable—such as during accidents, extreme weather events, or rare driving behaviors [6]. Traditional models often rely heavily on historical data and assume regularity in traffic flow, which limits their ability to generalize in unpredictable or abnormal conditions. Moreover, they lack the ability to generate new, synthetic data to supplement real-world inputs, which restricts their effectiveness in training autonomous driving systems or simulating complex traffic scenarios.

The development of AI, and especially generative AI, has brought about a radical change in the way infrastructure is designed. Algorithms that can produce new outputs—like text, graphics, or, in our case, design layout—based on patterns and data gathered are known as generative AI.

The latest advancements in generative AI have opened up new avenues of infrastructure planning. Generative AI models such as Generative Adversarial Networks (GANs) and Variational Autoencoders (VAEs) can learn from enormous datasets and generate new, context-dependent design alternatives [7]. Such models can generate road layouts, tune traffic configurations, and adapt designs to varying constraints such as geography, traffic demands, and safety norms. GANs generate artificial traffic environments, which assist in better training autonomous vehicles. Auto-encoders and Graph Neural Networks are applied to detect anomalies and model road networks, while Deep Reinforcement Learning (DRL) aids dynamic traffic light control and autonomous driving decision-making. These technologies complement each other to enable transportation systems to become smarter, predictive, and capable of coping with real-time traffic dynamics. By combining deep learning–based security models [8] with Digital Twins (DTs) [9], recent developments in Cooperative Intelligent Transportation Systems (CITS) have demonstrated encouraging outcomes. One such method builds a hybrid model that forecasts traffic circumstances and improves system security in real time by combining Convolutional Neural Networks (CNNs) with Support Vector Regression (SVR) [10]. In addition to increasing traffic prediction accuracy, this technique facilitates improved path design, quicker emergency response, and lower emissions and congestion. The model's use of virtual copies of actual traffic systems enables safe testing and optimization without interfering with real-time traffic. The findings point to a significant advancement in the development of transportation networks that are safer, more adaptable, and more sustainable.

Generative AI can be used as a design tool and a co-creator to produce scalable and adaptive solutions when combined with traffic simulation softwares and reinforcement learning techniques. Generative AI can rapidly generate a vast number of highly distinct design alternatives. Generative AI may simultaneously optimize for several, typically contradictory, goals by including various design elements and restrictions into its algorithms. Generative AI may significantly reduce design time and expenses by automating a large portion of the process, which will accelerate project timelines from conception to completion.

It can assist in producing more sustainable designs with less of an impact on the environment and more robust infrastructure that can withstand future adversities by exploring a wider range of design options and optimizing for environmental variables. Therefore, generative AI is a powerful tool for ideation through new design prospects, upending conventional design, and encouraging a creative culture in the civil engineering industry.

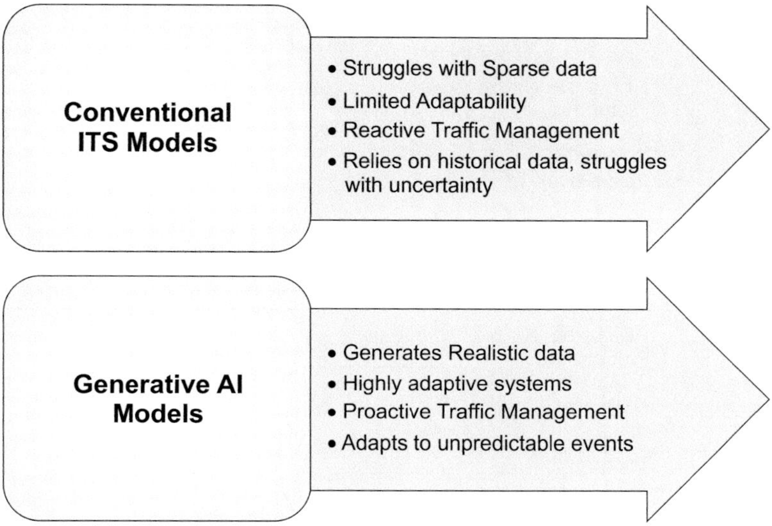

Fig. 8.3 Conventional ITS models vs Generative AI models

As shown in Figure 8.3, differing from traditional models, generative models like GANs, VAEs , and Diffusion Models are capable of learning the distribution of traffic environments and generating highly realistic synthetic data. This enables ITS frameworks to model infrequent but critical incidents (e.g., near-collisions or pedestrian jaywalking), complete missing sensor data, and facilitate the training of stronger and more adaptive systems. Generative AI also improves simulation environments and traffic digital twins to make it possible to move beyond reactive control toward proactive and predictive traffic management. This is a key advancement necessary in the creation of safe, autonomous, and human-aware transportation systems.

1.6 Key Technologies Powering Generative AI in ITS

Several strong models that are each intended to learn intricate data patterns and produce realistic, high-quality outputs are the foundation of generative AI in ITS and are depicted in Figure 8.4. These technologies allow systems to better train autonomous systems, forecast vehicle behavior, fill in missing sensor data, and simulate traffic situations.

- *Variational Autoencoders (VAEs)*: These models learn hidden patterns from data and can generate synthetic traffic data by reconstructing it from compressed representations. They are especially useful for handling uncertainty in traffic environments.

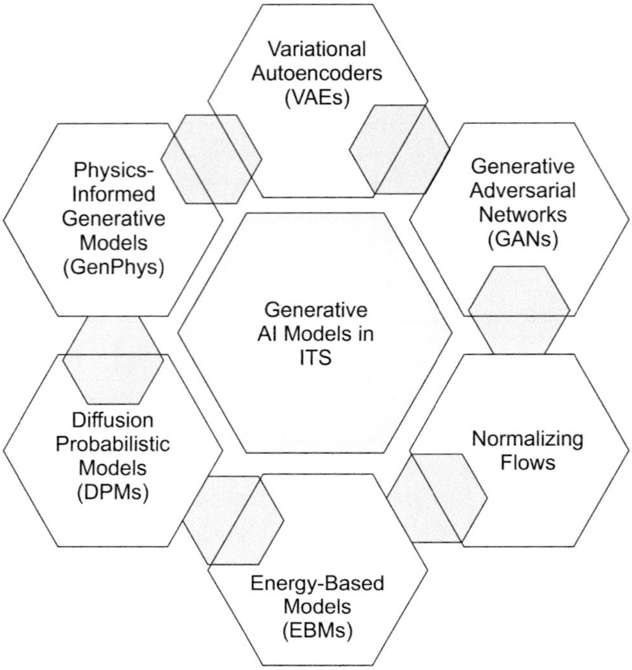

Fig. 8.4 Technologies powering generative AI in ITS

- *Generative Adversarial Networks (GANs)*: GANs use a competitive learning process between two models—a generator and a discriminator—to produce highly realistic data, such as images of traffic scenes or accident scenarios. They are great for simulating complex driving behavior.
- *Normalizing Flows*: These techniques allow for transforming simple data distributions into complex ones. In ITS, they help estimate the probability of rare traffic events, making them useful for anomaly detection and traffic forecasting.
- *Energy-based Models (EBMs)*: These models assign energy scores to data, helping systems distinguish realistic traffic conditions from unrealistic ones. They are helpful in modeling multivariable traffic patterns.
- *Diffusion Probabilistic Models (DPMs)*: DPMs generate data by gradually refining noise into meaningful outputs. They are excellent for simulating complex traffic conditions in a step-by-step, controlled way.
- *Physics-Informed Generative Models (GenPhys)*: These models integrate physical principles (such as vehicle motion or traffic flow equations) into generative AI, guaranteeing that simulations represent real-world restrictions and behaviors.

2. Related Work

Momeni et al. [10] utilized an easy-to-use, logic-based language to store rules for the design of highway passing lanes, which were automatically tested and created using the BIM-kit software environment. The findings revealed dramatic enhancements in rule compliance from 61.82% prior to implementing generative design methods to 91.31% following implementation. Such results identify the success of generative design as a powerful tool for engineers to develop compliant and innovative solutions for highway projects. The work of Li et al. [11] employs a new pipeline involving transformer-based GAN architectures in order to generate road network graphs with some predetermined attributes. The method addresses both the spatial and topological features and outperforms current image-based and graph-based techniques. Ye et al. [12] introduce a road network generation algorithm using Conditional Generative Adversarial Networks (CGANs) tailored to autonomous driving (AD) conditions. To measure the quality of the generated networks, the authors proposed a well-rounded framework that includes both static and dynamic test criteria, filling an important void in current methods, which tend to neglect dynamic testing. Through carefully designed experiments, the feasibility and effectiveness of the CGAN-based approach were validated, demonstrating that the generated road networks can enhance travel experiences, reduce energy consumption, and promote urban sustainability. Xie et al. [13] have proposed a hierarchical vehicular Generative Artificial Intelligence (GAI) scheme intended to improve Edge Intelligence (EI) for Internet of Vehicles (IoV) applications. The core approach is a collaborative inference mechanism that is distributed across RSUs and vehicles, enabling effective content creation with reduced latency and cloud-based GAI service-related security threats. One key problem discussed in the study is the local and shared inference splitting problem, which impacts both the latency of GAI services and their ability to generate content. To address it, the authors suggest using a backward induction-based algorithm based on a simple threshold-based policy for making inference splitting decisions. Gaier et al. [14] showed a method that solves issues in engineering generative models, namely in architectural design, through optimization, constraint satisfaction, and language models. It applies Quality-Diversity (QD) to generate a high-performing and diverse dataset. This dataset is utilized to fine-tune a language model for generating high-level design, which is then augmented into detailed, constraint-satisfying layouts with the use of the Wave Function Collapse algorithm. Dikshit et al. [15] studied the use of AI for improving vehicle routing and reducing urban traffic congestion. The methodology is based on an in-depth analysis of available AI algorithms, vehicle routing methods, and traffic management models. It combines real-time traffic information, road network features, and personal travel behavior to create smart routing policies. The suggested AI system uses machine learning and optimization techniques to learn from dynamic traffic conditions, determine the best routes, and reassign traffic streams to eliminate congestion pockets. Prakash et al. [16] presented an AI-based approach to route forecasting in multimodal transport networks, combining for the first time Monte Carlo simulations and

LSTM (Long Short-Term Memory) networks. The approach consists of creating a robust and diverse training dataset by using Monte Carlo simulations, as this reflects the stochastic nature of transport networks. Han et al. [17] explored the concept of foundation intelligence for smart infrastructure services in the context of Transportation 5.0. The methods involve discussing a suite of emerging technologies crucial for this intelligence, including digital twinning, parallel intelligence, large vision-language models, traffic simulation and transportation systems modeling, V2X connectivity, and decentralized/distributed systems.

Pachouri et al. [18] examined the upgradation of road-based technology to ensure safer mobility by incorporating AI, Big Data Analytics (BDA), and Augmented Reality (AR) to improve the road environment and enhance the development of digital road architecture. The paper explains different applications of AI, BDA, and AR, citing them as a new strategy toward ITS. Cuddapah et al. [19] introduced a new traffic control system known as Enhanced Transportation Technologies (ETT) to relieve congestion at an intersection that sees heavy traffic, in place of an inefficient older system. The system utilizes intelligent sensing such as vision sensors such as vision sensors (e.g., cameras), to surpass the limitations of traditional dynamic controllers and estimate traffic volumes. Machine learning techniques based on AI are utilized in order to allow machines to learn and observe the world around them through constant sensing and collaborating with ITS to enhance city planning, crowd movement, and traffic prediction. Large simulations proved that the developed model considerably decreases waiting time and enhances the average movement speed, performing better than controllers that utilize traditional sensors.

3. Framework for Road Infrastructure Design

The proposed framework shown below in Figure 8.5 follows a structured, data-driven, and iterative methodology that integrates generative modeling, traffic simulation, and reinforcement-based optimization for the design of intelligent road infrastructure. As shown in Figure 8.5, the process can be systematically divided into various key stages.

3.1 Data Collection and Integration

The process starts with integrating various datasets pertinent to urban mobility and spatial planning. Some of the datasets collected include historical and real-time traffic (via loop detectors, sensors, or GPS), GIS layers, satellite imagery, land-use zoning maps, and accident reports [20, 21, 22, 23, 24]. The disparate datasets were integrated into a shared geospatial framework for analysis. The collected datasets often differ in terms of format, spatial reference systems, scale, and granularity. A key challenge is merging these heterogeneous sources into a coherent, unified framework. A successful transportation system design starts with a strong base of varied, high-quality data. This complete dataset is the backbone for generative AI model training and testing that reproduce realistic city transportation scenarios.

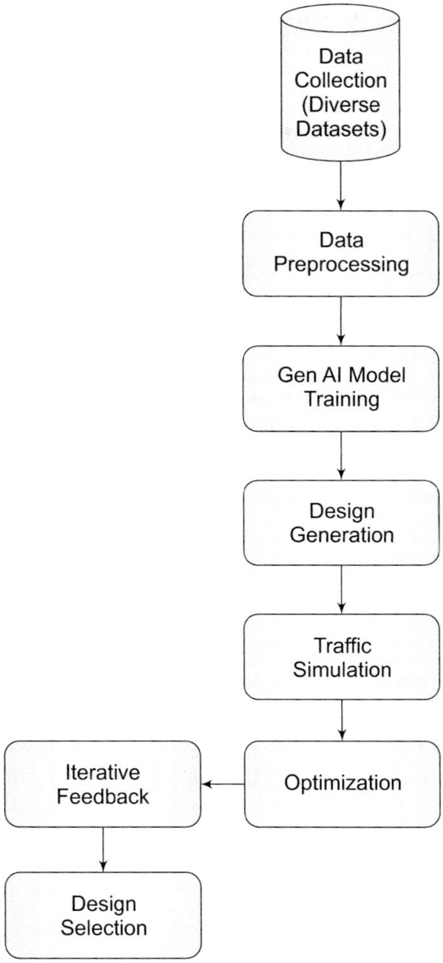

Fig. 8.5 Flowchart for road infrastructure design

3.2 Data Preprocessing and Feature Engineering

The effectiveness of any AI or machine learning system is significantly influenced by the quality and representation of the input data. Once the raw data are collected from heterogeneous sources, including traffic logs, satellite imagery, GIS layers, and zoning maps, they undergo a rigorous preprocessing and feature engineering phase to ensure that they are clean, consistent, and machine-readable. This structured and standardized dataset forms the foundation for training the generative AI model, which later simulates traffic scenarios and designs.

3.2.1 Data Cleaning and Normalization

Real-world urban and traffic datasets often suffer from issues such as the following: missing values, inconsistent naming conventions, outliers and anomalies, and mixed data types (e.g., numerical, categorical, geospatial). To address these issues, a combination of automated scripts and manual verification was applied. The key steps include:

- *Handling Missing Values*: Techniques such as mean/mode imputation, forward fill, or model-based interpolation (e.g., KNN-K nearest neighbour imputer) are used.
- *Outlier Detection*: Z-score or IQR (Interquartile Range)-based filtering is applied to continuous features, such as average vehicle speed or traffic volume.
- *Normalization*: Features are scaled to standard ranges using Min-Max Scaling or Standardization (z-score) to ensure model stability.
- *Unit Harmonization*: Measurement units are converted consistently (e.g., meters vs kilometers, km/h vs m/s).

3.2.2 Feature Encoding

The data may include categorical variables, such as road type (e.g., arterial, highway, local), land use (residential, industrial, commercial), traffic signal types, intersection control, or lane marking. These non-numeric attributes were transformed into machine-readable formats using encoding techniques. These encoded features are crucial for helping the generative AI model learn the structural and semantic distinctions between different road types or land-use areas.

3.2.3 Geospatial Data Transformation

Because much of the ITS data are spatially anchored, they must be processed into a format suitable for AI modeling. Geospatial processing techniques such as Coordinate Transformation, Vector to Raster Conversion, and Spatial Tensor Creation are used. Spatial features are encoded as tensors where each cell may represent features such as: traffic density, road width, presence of intersections or roundabouts, adjacent land use zoning. These tensors serve as inputs for CNN-based generative models or diffusion models capable of learning spatial dependencies.

3.2.4 Temporal Feature Engineering

Urban traffic patterns exhibit strong temporal dynamics such as morning peaks, weekend trends, or seasonal variation. Therefore, timestamps are transformed into meaningful cyclic or trend-aware features as shown in Table 8.1. This step enables the model to understand temporal cycles inherent in traffic flow and infrastructure usage.

Table 8.1 Temporal features

Raw Feature	*Engineered Feature*
Timestamp	Hour of day, Day of week
Date	Weekend flag, Holiday flag
Time series	Rolling average traffic volume

3.2.5 Dataset Structuring and Storage

Once all features are prepared, the final dataset is: Indexed by geospatial and temporal keys, Stored in efficient formats like .parquet or .hdf5 for faster I/O, and Partitioned into training, validation, and testing sets (e.g., 70:15:15 split). This clean and well-structured dataset feeds directly into the next stage generative AI model training.

3.3 Training of Generative Model

In order to facilitate the automatic generation of road infrastructure designs, a deep generative model is trained to discover the intricate spatial and contextual relationships that exist in current road networks. For this, sophisticated generative architectures like Conditional Generative Adversarial Networks (cGANs) and VAEs are utilized. These models are particularly suitable to capture the distribution of structured data and generate realistic new designs that follow both learned patterns and defined constraints.

The training is initiated through the preparation of an inclusive dataset with real-world road topologies as well as corresponding contextual characteristics. Road network data tends to be collected from databases like OpenStreetMap, municipal GIS, or satellite imagery. They are converted into workable forms like grid-based images, graphs, or sequences of coordinates. Contextual features that have an influence on road planning—such as traffic volume, zoning policy, population dispersion, terrain elevation, location of accidents, and proximity to strategic infrastructure—are extracted and encoded to serve as conditional inputs of the model. For instance, accident black spots can be represented in heatmaps, zoning details in category maps, and traffic volume as temporal or spatial patterns. These features help guide the model to learn how different real-world constraints affect road layout decisions.

During training, techniques such as early stopping, regularization, and data augmentation are utilized to improve generalization and avoid overfitting. The model is quantitatively evaluated using metrics such as Intersection over Union (IoU), Structural Similarity Index (SSIM), and Fréchet Inception Distance (FID) to measure the similarity of the generated road layouts to actual-world counterparts. Qualitative evaluation by urban planning experts is also carried out to assess the functional and regulatory feasibility of the generated road plans.

After it has been trained, the model can be used to generate new road designs for various planning scenarios. The planners, for example, can input specific constraints, such as areas of heavy traffic or limited lands, and the model will provide feasible road designs that satisfy such specifications. This capability facilitates rapid prototyping of infrastructure, simulation of scenarios, and optimization during the early phases of planning. Overall, the generative model is a good means of introducing AI-supported decision-making into the transport and urban planning process.

3.4 Design Generation

Once it has been properly trained, the generative model proceeds into the design generation process, where it generates novel road layout configurations specific to the input constraints and contextual parameters presented. The generated designs are not arbitrary but are data-driven, reasonable configurations that encapsulate the patterns learned during real-world road networks. Every produced layout contains necessary elements like intersections, road arrangements, road hierarchies (highways, arterials, local streets), and connectivity structures that provide navigability and functional movement in the planned infrastructure. The generative model considers spatial continuity and urban planning philosophy by aligning roads with zoning needs, reducing conflict points, and providing adequate access to nearby land parcels.

The output of this step is formatted in machine-readable, vector-based output like ESRI Shapefiles or OpenStreetMap (OSM) XML, which maintains geospatial integrity and allows for easy integration with GIS and traffic simulation software. Such formats support explicit representation of road geometry, attributes (e.g., lane count, speed limits), and topology, which are required inputs to downstream applications such as transportation modeling, safety analysis, and infrastructure optimization. Additionally, compatibility with simulation software like SUMO (Simulation of Urban MObility), MATSim (Multi-Agent Transport Simulation), or AIMSUN facilitates dynamic analysis of the created layouts under different traffic conditions, pedestrian interactions, and multimodal transport needs.

3.5 Traffic Simulation and Evaluation

After generating candidate road layout designs, the next critical phase involves evaluating their real-world effectiveness and operational feasibility through traffic simulation. This step ensures that the proposed infrastructure not only adheres to design standards but also performs efficiently under dynamic traffic conditions. To achieve this, the generated layouts are imported into microscopic or agent-based traffic simulators such as SUMO or MATSim. These tools offer high-fidelity modeling of individual vehicle and pedestrian behavior, enabling detailed analysis of how traffic flows through the designed network over time.

The results from the simulations guide comparative analysis among multiple generated designs, highlighting trade-offs among efficiency, safety, and sustainability. This systematic evaluation not only supports data-driven decision-making in road infrastructure planning but also enables iterative refinement of designs. Based on key performance indicators' (KPI) feedback, the generative model may be reconditioned or tweaked to emphasize specific outcomes (e.g., decreased emissions or enhanced safety), making the entire workflow adaptable and optimization-ready.

3.6 Iterative Feedback Loop

The penultimate phase of the generative design approach is the incorporation of an iterative feedback loop, which is instrumental in the optimization of road infrastructure configurations. This process-oriented feedback turns the system into a closed-loop structure, wherein feedback from the simulation environment is used to drive future design iterations. Following the first generation of road layout proposals and evaluation with traffic simulation tools like SUMO or MATSim, important performance indicators (KPIs)—such as travel time, vehicle delay, emissions, safety, and throughput—get harvested and analyzed. The KPIs are quantitative feedback indicators that show how well the generated designs perform as well as their real-world feasibility.

This feedback is then formally fed back into the generative model, enabling it to refine its parameters, modify its design strategies, and learn from the success and failure of past outputs. This might involve reinforcement learning mechanisms, in which case the model learns through a reward signal of KPI performance, or gradient-based fine-tuning, in which case loss functions are optimized to promote more desirable results (e.g., less congestion or greater safety). The model uses this feedback not only to avoid repeating suboptimal design choices but also to progressively refine its internal representation of what constitutes an efficient, safe, and regulation-compliant road network.

This feedback mechanism creates a co-design platform in which human planners and AI collaboratively work between iterations, domain experts can step in and give constraints or preferences, and the model integrates them into subsequent generations. Finally, the closed-loop nature renders the generative system not just adaptive and self-correcting but also resilient in handling the composite and dynamic character of contemporary transport planning problems.

3.7 Design Selection and Visualization

The last step is the selection and visualization of the optimized designs generated through iterative feedback and simulation assessment. At this point, a curated set of the most promising road designs is assembled, each having demonstrated strong performance across a range of KPIs such as average travel time, congestion levels, fuel consumption, emissions, safety scores, and pedestrian accessibility. These designs are systematically ranked or scored using a multi-criteria decision analysis (MCDA) framework or a weighted KPI aggregation method, depending on the planning objectives and stakeholder priorities.

To support informed decision-making, the results are presented to urban planners, transportation engineers, and policymakers via a visualization dashboard. This dashboard provides an interactive, user-friendly interface where users can explore the spatial configuration of each layout alongside its corresponding performance metrics. Visual components may include geospatial maps of the road networks, heat maps of traffic density or delay, bar charts comparing KPI scores, and simulation playback features to observe dynamic vehicle and pedestrian flows. Additional filters and sliders can allow users to adjust weighting factors for KPIs, simulate "what-if" scenarios, or narrow down designs based on specific constraints (e.g., environmental impact or budget limits).

This stage bridges the gap between algorithmic generation and human-centric implementation, ensuring that the final selected road layout aligns with both technical optimization and real-world planning goals. It enhances trust in AI-generated infrastructure solutions by grounding them in measurable outcomes and making them interpretable and actionable for decision-makers. This systematic methodology not only enables the automatic generation of road networks but also ensures that designs are continuously improved based on performance feedback and real-world constraints.

4. Conclusion

The chapter demonstrated that leveraging generative AI and reinforcement learning presents a ground-breaking methodology for the design and optimization of urban road infrastructure. Unlike conventional approaches constrained by manual processes and limited design space exploration, the proposed data-driven and iterative framework harnesses advanced generative models to produce adaptive, multi-objective, and context-sensitive road layouts. Integration with ITS and real-time traffic simulation enables robust evaluation and continuous improvement based on quantifiable performance metrics such as efficiency, safety, and environmental impact. Incorporating stakeholder input via visualization dashboards ensures that technical innovation aligns with real-world planning goals, enhancing transparency and decision-making. The experimental results confirm that generative AI not only increases the speed and resilience of infrastructure design but also empowers cities to address future challenges like congestion, sustainability, and evolving mobility patterns. Ultimately, this approach signifies a shift toward intelligent, sustainable, and collaborative infrastructure planning, laying the foundation for next generation urban mobility and smart city development.

References

[1] Creß, C., Bing, Z. and Knoll, A.C. (2024). Intelligent transportation systems using roadside infrastructure: A literature survey. *IEEE Transactions on Intelligent Transportation Systems*, *25*(7), 6309–6327.

[2] Kulkarni, D.S., Deshpande, A.B., Bhandurge, S.A., Estrela, V.V. and Deshpande, A. (2025). Impact of textual length on lexicon-based sentiment classification for Hindi reviews. *Cureus Journal of Computer Science*, *2*. DOI https://doi.org/10.7759/s44389-025-05622-w

[3] Kulkarni, D.S. and Rodd, S.F. (2022). Towards enhancement of the lexicon approach for Hindi sentiment analysis. In *IOT with Smart Systems: Proceedings of ICTIS 2021*, Tomonobu Senjyu, Parakshit Mahalle, Thinagaran Perumal, Amit Joshi *2*, 445–451. Singapore: Springer Nature Singapore.

[4] Patil, A.A. and Kulkarni, D. (2015). A survey on: secure data deduplication on hybrid cloud storage architecture. *International Journal of Computer Applications*, *110*(3), 29–32.

[5] Wang, F.-Y. et al. (2023). Transportation 5.0: The DAO to safe, secure, and sustainable intelligent transportation systems. *IEEE Transactions on Intelligent Transportation Systems*, *24*(10), 10262–10278.

[6] Yan, H. and Li, Y. (2025). Generative AI for intelligent transportation systems: Road transportation perspective. *ACM Computing Surveys*, *57*(12), 315.

[7] Xu, H., Omitaomu, F., Sabri, S. et al. (2024). Leveraging generative AI for urban digital twins: A scoping review on the autonomous generation of urban data, scenarios, designs, and 3D city models for smart city advancement. *Urban Informatics*, *3*, 29.

[8] Estrela, V.V., Deshpande, A., Sroufe, R., Lopes, R.T., Grata, E.G.H., Andreopoulos, N., Terziev, A. and Laghari, A. A. (2023). Digital twin framework for intelligent healthcare facilities through ISO/IEEE 11073. In *Intelligent Healthcare Systems*, ed. V. V. Estrela, 279–297. CRC Press.

[9] Estrela, V.V., Bukhsh, K., Malook Rind, M., Shaikh, S., Khan, A.A., Laghari, A.A., Deshpande, A., Andreopoulos, N. and Terziev, A. (2023). Deep learning as a driving force for better drug development. In *Intelligent Healthcare Systems*, ed. V. V. Estrela, 217–235. CRC Press.

[10] Momeni Rad, F., Sydora, C. and El-Basyouny, K. (2024). Leveraging generative design and point cloud data to improve conformance to passing lane layout. *Sensors*, *24*(2), 318.

[11] Li, T., Li, R., Ye, S., Zhang, Z., Yin, Z., Li, S. and Pan, Z. (2024). CRTGAN: Controllable road network graphs generation via transformer based GAN. *2024 International Joint Conference on Neural Networks (IJCNN)*, *33*, 1–8.

[12] Ye, S., Li, T., Yang, G., Lv, P., Li, H. and Pan, Z. (2024). Autonomous driving road network generation and evaluation based on generative artificial intelligence. *Proceedings SPIE 13018, International Conference on Smart Transportation and City Engineering (STCE 2023)*, 130183I.

[13] Xie, G., Xie, R., Zhang, X., Nie, J., Tang, Q. and Niyato, D. (2024). Enhancing vehicular edge intelligence through distributed collaborative generative AI inference. *ICC 2024–IEEE International Conference on Communications*, 4560–4565..

[14] Gaier, A., Stoddart, J., Villaggi, L. and Sudhakaran, S. (2024). Generative design through quality-diversity data synthesis and language models. *GECCO '24: Proceedings of the Genetic and Evolutionary Computation Conference*, 823–831.

[15] Dikshit, S., Atiq, A., Shahid, M., Dwivedi, V. and Thusu, A. (2023). The use of artificial intelligence to optimize the routing of vehicles and reduce traffic congestion in urban areas. *EAI Endorsed Transactions on Energy Web*, *10*. Available at https://publications.eai.eu/index.php/ew/article/view/4613 (accessed on 29 December 2025).

[16] Prakash, S., Mehta, U. and Sharma, B. (2024). Artificial intelligence-driven multimodal route planning: Addressing dynamic unavailability and disruptions. *IEEE Access*, *12*, 172088–172100.

[17] Han, X., Meng, Z., Xia, X., Liao, X., He, Y., Zheng, Z., Wang, Y., Xiang, H., Zhou, Z.,Gao, L., Fan, L., Li, Y. and Ma, J. (2024). Foundation intelligence for smart infrastructure services in transportation 5.0. *IEEE Transactions on Intelligent Vehicles*, *9*(1), 39–47.

[18] Pachouri, V., Kathuria, S., Gehlot, A., Negi, P., Kathuria, A. and Chaudhary, M. (2023). Automation of road infrastructure: Indulgence of AI, big data analytics & AR. *2023 3rd International Conference on Pervasive Computing and Social Networking (ICPCSN)*, 512–517.

[19] Cuddapah, A., Sharma, S., Nassa, V.K., Agrawal, S.K., Rajasekaran, A. and Mahaveerakannan, R. (2024). Artificial intelligence powered congestion free transportation system through extensive simulations. *Journal of Machine and Computing*, *4*(1), 250–260.

[20] Caron, F., Duflos, E., Pomorski, D. and Vanheeghe, P. (2006). GPS/IMU data fusion using multisensor Kalman filtering: Introduction of contextual aspects. *Information Fusion*, *7*, 221–230.

[21] Estrela, V.V., Khan, A.A., Shaikh, A.A., Laghari, A.A., Dootio, M.A., Hussain, M., Jumani, A.K. and Ayub, R. (2023). Some issues regarding content-based image retrieval (CBIR) for remote healthcare theradiagnosis. In *Intelligent Healthcare Systems*, V.V. Estrela, 110–134. CRC Press.

[22] Akhavi Zadegan, A., Vivet, D. and Hadachi, A. (2024). DELTA: Integrating multimodal sensing with micromobility for enhanced sidewalk and pedestrian route understanding. *Sensors*, *24*(12), 3863.

[23] Bostanci, E., Bostanci, B., Kanwal, N., Clark, A.F. (2018). Sensor fusion of camera, GPS and IMU using fuzzy adaptive multiple motion models. *Soft Computing*, *22*, 2619–2632.

[24] Paus, T. and Kum, H. C. (2024). *Digital Ethology: Human Behavior in Geospatial Context*. Volume 33. MIT Press.

9 Neural Network–driven Driver Wellness Analytics: Predicting Mental Fatigue through Whole-Body Vibration Patterns and Psychosocial Health Monitoring

Gourav Vivek Kulkarni[1*] and Samiksha Shivaji Mayekar[2]

In today's fast-paced, technology-driven society, mental fatigue among vehicle operators has emerged as a critical safety and public health concern. Health being a combination of physical, mental, and social well-being, it has become essential to address the societal aspect of the drivers of a nation's economy. With growing dependence on road-based transport and extended working hours, drivers are increasingly exposed to physical stressors such as whole-body vibrations (WBVs), noise, and cognitive load, which cumulatively contribute to fatigue, burnout, and impaired decision-making. This chapter explores the multifaceted impact of vibration-induced fatigue, particularly its psychosocial and physiological dimensions, drawing from recent empirical studies and quantified findings. Moving beyond traditional fixed rest models, the chapter critically evaluates adaptive fatigue management strategies enabled by wearables, machine learning, and dynamic scheduling. It presents case studies and frameworks that shift the paradigm from reactive to predictive fatigue mitigation. Further, it investigates the role of advanced neural network architectures, including Convolutional Neural Networks (CNNs), Long Short-Term Memory (LSTM), and hybrid models, for processing biosignals and behavioral data in real time. Through the perspective of edge computing and artificial intelligence (AI) optimization algorithms, the chapter highlights the importance of scalable, responsive, and context-aware systems for human and machine performance monitoring. The final sections consider condition-based maintenance in mechanical systems and advocate for multimodal sensor integration to enhance detection accuracy. The chapter concludes with a forward-looking view on privacy-preserving AI, digital twins, and the integration of driver fatigue detection with Intelligent Transportation Systems, underlining the need for interdisciplinary solutions to safeguard driver health and to ensure public safety.

1. Introduction

Whole-body vibration (WBV) largely affects driver performance, contributing to drowsiness, fatigue, musculoskeletal issues, and reduced interaction efficiency

[1,2] Department of Mechanical Engineering, KLS Gogte Institute of Technology, Belagavi, Karnataka, India.

* Corresponding author: gvkulkarni@git.edu

DOI: 10.1201/9781003648147-09

with in-vehicle systems. Recent research integrates artificial intelligence (AI) and Artificial Neural Networks (ANNs) to better predict and manage these impacts, offering innovative solutions for real-time monitoring and long-term health assessment [1, 2, 3, 4]. This review compares studies examining WBV's cognitive, physical, and ergonomic effects, highlighting how AI-driven approaches can effectively help overcome associated risks. The studies considered present a multidimensional perspective on how human factors, vibration exposure, technological systems, and ergonomics converge in transportation and occupational safety, and how they interact with each other and contribute to overall human health and safety.

The main idea is a central WBV concern and its diverse effects on human health and performance. A review by Bhuiyan et al. [1] focus on the acute cognitive effects on drivers of WBV, primarily drowsiness, which contributes significantly to road accidents. It shows that low-frequency WBV in the 4–10 Hz range can induce substantial drowsiness within just 30 minutes of exposure. This finding is crucial because it identifies a preventable hazard that can be overcome through changes in vehicle design by modifications and the addition of more effective vibration isolation. However, Aguilar et al. [2] address WBV from a chronic health standpoint, specifically musculoskeletal disorders. The scope of concern thus further shifts from immediate safety (drowsiness and accidents) to long-term worker health. The study in [2] uses ANNs to assess compressive forces on the lumbar spine, thereby adding a predictive and personalized layer that is not generally considered in recent literature.

An interesting thing is that the harmful impact of WBV is well recognized, but the consequences are different, namely, cognitive fatigue versus physical degradation occurring over various timescales. Yet, neither fully explores how these effects could affect real-world driving scenarios where short-term drowsiness may lead to unsafe operation. Long-term spinal damage accumulates silently, signaling a research gap: integrated assessment frameworks that account for both immediate and cumulative health effects using AI [3, 4, 5, 6, 7].

Another survey on driver safety systems [8] provides a picture of recent advancements in these systems. While WBV is not the only focus, the survey identifies fatigue and inattention as major causes of accidents. It explores how emerging technologies like the Internet of Things (IoT) and machine learning (ML) can detect and prevent such conditions. However, this study fails to integrate WBV as a specific variable in the safety solutions. It indeed needs to be considered as a strong causal link between WBV and driver drowsiness has been highlighted earlier [1]. Bridging this gap would involve integrating WBV data into existing driver monitoring systems, perhaps using wearable sensors or embedded seat- or floor-level measurements to feed ML-based detection algorithms [9, 10, 11]. This concept is hinted at but not fully explored in recent literature.

For a complete approach to vibration-related risks, future research must merge these perspectives into unified concepts that assess immediate, chronic, and interactive effects of WBV. These frameworks should utilize ML not just for predictive modeling but also for adaptive interface and vehicle system responses [8, 9]. Ethical considerations and privacy safeguards must be the basis for all such efforts, drawing from the AI-focused insights. Through such interdisciplinary integration, safer and healthier systems can be developed. These would also be fair, ethical, and technologically robust.

2. Mental Fatigue Induced by Vibrations

2.1 Psychosocial Health and Vibration Exposure

Exposure to WBV and noise presents a number of challenges to driver well-being, particularly through physiological stress mechanisms, mental fatigue progression, and psychosocial strain. The studies analyed agree that WBV is not merely a physical discomfort but also contributes to deeper physiological and psychological outcomes. These include occupational fatigue and burnout. A study [12] highlights the physiological importance of WBV's effects by quantifying drivers' exposure to vibration (0.99 m/s^2) and noise (82.18 dB). These were correlated with fatigue using a standardized Swedish fatigue questionnaire. The results clearly demonstrated a significant association between WBV and fatigue, underscoring the role of physical factors, such as seat type and vehicle condition, as mediators of these physiological stressors. The finding that fatigue levels rise in correlation with vibration exposure illustrates the biological strain imposed on the body's neuromuscular and sensory systems, further exacerbating exhaustion, even with short- to medium-duration exposures [12]. In contrast, another study [13] broadens the scope by linking WBV exposure with complex mental fatigue patterns, psychosocial responses, and job burnout. Using the Maslach Burnout Inventory, the research reveals that bus drivers show medium to high levels of emotional exhaustion, depersonalization, and lack of personal accomplishment.

The investigation in [12] implicitly links dissatisfaction and occupational fatigue to vehicle ergonomics, whereas [13] explicitly associates high burnout scores with vibrational exposure and work overload. The collective stress from vibration, noise, and the job is likely to lead to emotional strain, which is expected to alter drivers' long-term engagement with their work, contributing to reduced job satisfaction and increased turnover risk.

2.2 Traditional vs. Predictive Rest Management

Fixed-interval rest periods were considered a standard in various work environments. These are increasingly being challenged by emerging research and technologies that suggest dynamic, personalized scheduling based on real-time data and individual needs. The limitations of static rest intervals are explained in both human-centered and industrial applications. The work in [14] shows the inefficiency of collecting occupant comfort data at fixed intervals, leading to redundant feedback and user fatigue. Instead, it promotes a scenario-based, optimized sampling method using smartwatches and Graph Neural Networks (GNNs) [15, 16]. This allows dynamic triggers for feedback collection, clearly showing a case against fixed schedules. Uniform intervals fail to account for context-specific needs or changing user states, leading to inefficiencies and reduced engagement.

In an industrial scenario, the same rigidity of fixed schedules applies to job shop environments, where the unpredictability of machine operations and maintenance needs makes static scheduling less optimal. Another study [17] introduces an adaptive, automated scheduling model that incorporates real-time data and earned-value project management principles. This model reduces unnecessary rescheduling by dynamically adjusting processing times and due dates. Like worker rest periods, machine operations cannot thrive under static planning [16]. Adaptive, data-informed rescheduling minimizes downtime and improves overall system efficiency. This reinforces the broader theme that dynamic approaches are better than fixed frameworks. Similarly, a further study [18] presents a scheduling problem in semiconductor manufacturing that allows for flexible maintenance intervals. The research acknowledges the complex, Non-deterministic Polynomial-time (NP)-hard nature of such problems and introduces the Improved Self-Adaptive Harmony Search algorithm.

The authors of [19] extend this dynamic personalization concept to human-centric systems through Human Digital Twins (HDTs). HDTs represent a shift from static medical check-ups and generalized health recommendations to continuously updated, individualized physiological and behavioral data streams. HDTs can optimize not only healthcare delivery but also real-time rest and recovery needs. This suggests that individual rest schedules based on real physiological data could become feasible. Just as industrial digital twins [20] predict machine failures, HDTs could predict fatigue accumulation or cognitive overload in workers [21, 22]. This could enable rest periods tailored to when the individual actually needs them, rather than when the clock dictates.

These investigations suggest discontinuing fixed-interval rest models and adopting dynamic, personalized systems. Whether in human labor or machine operation, rigid intervals ignore the variability that naturally occurs in systems and individuals. Adaptive scheduling, empowered by AI, ML, digital twins, and context-aware algorithms, offers scalable, efficient, and more humane solutions. Fixed intervals may provide simplicity, but the issue is often inefficiency, disengagement, and even physical strain. Personalized rest scheduling powered by AI would ensure recovery or maintenance efforts with real-time needs, leading to better performance, health outcomes, and resource optimization. As demonstrated across these papers, the future of scheduling would be dynamic, context-sensitive models for both humans and machines.

3. ANN Architectures for Predictions

3.1 Multimodal Feature Integration Models

The emergence of deep learning has revolutionized how temporal and physiological data are interpreted. The key domains include structural health monitoring (SHM), human activity recognition (HAR), and physiological signal processing. A common aspect across these studies is the increasing effectiveness of specialized neural network architectures, namely, Long Short-Term Memory (LSTM) networks for capturing temporal dependencies, Convolutional Neural Networks (CNNs) for spatial and frequency-domain pattern recognition, and hybrid models that leverage the strengths of both [23, 24].

Studies akin to [25, 26] demonstrate the power of hybrid ANN frameworks that combine CNNs and LSTMs to process raw sensor data streams. In [25], a time-distributed 1D CNN-LSTM model is used to analyze vibration data from structural systems. CNN layers extract spatial patterns from multisensor data, while LSTM layers model long-term temporal dependencies. This hybrid approach achieved extremely high damage localization accuracy (96.6%) and pattern recognition accuracy (99.6%), showcasing its potential in complex SHM tasks where temporal vibration patterns are critical.

Another study [26] introduces a parallel 2D CNN-LSTM model for HAR, where CNNs process spectrograms and LSTMs handle raw time-series data. By combining feature sets from both branches, the model achieves better results than baseline architectures, indicating that fusing spatial and temporal feature-extraction pathways yields a more comprehensive understanding of activity dynamics. A study [27] further supports the role of CNNs in robust physiological signal processing. By transforming noisy 1D accelerometer data into multiple 2D representations and feeding them into a multi-input CNN, the study overcomes the need for complex preprocessing. This approach not only enhances noise resilience but also improves accuracy and precision.

Together, these studies present the need for hybrid [28] and multimodal deep learning architectures [29]. LSTM networks excel at modeling temporal vibration patterns and long-term dependencies in sensor data. CNNs are highly effective in extracting nuanced spatial or frequency-domain features from physiological signals, especially when applied to transformed or multidimensional inputs.

3.2 Real-Time Processing Capabilities

The convergence of Edge Computing with optimization strategies for network efficiency is rapidly redefining the design of real-time and intelligent systems, particularly for IoT and mobile applications. Various studies [30, 31, 32] illustrate two central imperatives in this domain. The first is the need for immediate, low-latency responses via edge implementations, and the second is the need to optimize network parameters to balance computational accuracy with efficiency.

A study [30] highlights how this model reduces computation latency and energy consumption by supporting task offloading to nearby edge servers. It emphasizes that not all workloads benefit equally from offloading. The decisions must be context-aware, accounting for variables such as device capacity, wireless conditions, and task complexity, underscoring the need for intelligent, dynamic scheduling mechanisms.

Another study [31] expands on this by introducing a metaheuristic Artificial Bee Colony (ABC) optimization algorithm designed to manage workloads on edge servers under tight constraints of low network latency and high Quality of Service (QoS). There are essentially two pillars of immediate responsiveness in large-scale edge environments.

i. In the context of resource-constrained edge nodes and exponentially growing IoT traffic, striking a balance between accuracy and computational efficiency becomes vital. The work reviewed in [30] categorizes a range of optimization strategies.
ii. Another study [31] provides a broader systems-level perspective, discussing edge-computing-driven IoT (ECDriven-IoT) and its potential to outperform cloud-centric architectures. It identifies key architectural and algorithmic components. These include network communication layers to computation models. These must be optimized to avoid bottlenecks, reduce redundancy, and handle heterogeneous data flows in a scalable manner.

4. Vibration Pattern Analysis and Mental State Correlation

There are studies [33, 34] that focus on the automotive domain, particularly on drivers' mental and behavioral states. Initially, focus is on electroencephalography (EEG) as a core input to intelligent assisted driving systems (IADS). EEG-based techniques for detecting driver state and behavior consider data generation at various stages. This runs from the acquisition and enhancement of signals, through feature extraction and classification, to post-processing. In contrast, another study [34] goes beyond EEG and employs dynamic multi-biosignal fusion (DMBF) to detect driver and passenger mental states. The study adopts a quantitative approach, employing a confidence-aware learning mechanism and spatial-temporal attention to dynamically weigh biosignals based on signal quality. The model was tested on five datasets, yielding F1 scores of 0.5569 (motion sickness), 0.7187 (drowsiness), 0.6647 (distraction), 0.9378 (emotions), and 0.8092 (sustained attention). These results demonstrated superior performance compared to baseline methods, highlighting the robustness and generalizability of the DMBF model. While [33] provides an exhaustive conceptual and methodological overview, [34] offers concrete performance evidence, making it better for implementation.

Chen et al. [35] also focused on EEG-based driver state detection and the challenge of recognizing unfavorable driving states (UDS) using functional connectivity measures. They introduced a novel amplitude envelope correlation with leakage correction (AEC-c) approach in multiple frequency bands. By applying random subspace ensemble learning with k-nearest neighbors (KNN), the study achieved a classification accuracy of 96.88% [20]. Additionally, statistical analysis showed significantly lower AEC-c values for UDS compared to non-unfavorable driving states (NUDS), especially in beta and gamma bands.

Regarding mechanical systems, the condition monitoring of rolling element bearings (REBs) using acoustic emission (AE) signals was studied [36]. It proposed a semantic segmentation-based fault diagnosis framework. It includes preprocessing, fault fingerprint mapping, and a post-processing clustering step.

A workplace safety application [37] was introduced that uses IoT-based sensing and Support Vector Machines (SVMs) to detect drowsiness among hospital workers. Unlike the previous studies, which focused on real-time biosignal processing, this research takes an environmental and behavioral data approach via IoT infrastructure. The model is trained on comprehensive datasets that include ambient and physiological inputs and aims for real-time intervention.

5. ANN Implementation Strategies

The integration of dynamic rest scheduling, personalized wellness profiling, and predictive analytics into automotive and fleet systems presents a transformative approach to road safety, operational efficiency, and driver well-being. The reviewed studies contribute complementary insights that collectively shape the future of intelligent, health-aware transport ecosystems.

Ahlström et al. [38] critically explore dynamic rest scheduling in the context of highly automated vehicles. The focus is on whether in-cab rest or sleep during autonomous driving can be effectively utilized to extend allowable driving times. The proposed algorithm incorporates physiological and behavioral indicators such as heart rate and camera-based sleep detection, marking a shift from rigid schedules to adaptive, biofeedback-driven rest modeling. While this framework is conceptually robust, there exist some limitations.

In contrast, Oyebode et al. [39] emphasize the use of personalized wellness profiles in a meta-analysis of ML-based health systems. Its relevance to driver-specific fatigue modeling is significant. The paper suggests AI-driven personalization strategies to address variability in human responses to stress, fatigue, and environmental conditions.

The PREPIPE framework [40] introduces a predictive maintenance pipeline based on onboard sensor data, aiming to detect oxygen sensor degradation before failure. While it is not completely related to fatigue, the implications for integrating driver condition data are clear.

Considering fleet management [41], the study evaluates the effectiveness of a data-driven, telematics-informed safety and eco-driving training program. The study highlights the value of behavioral analytics in identifying risky patterns like harsh braking or excessive idling and improving them through feedback and training.

The application of ML to insurance risk assessment [42] offers a broader socio-behavioral view. Associating demographic, health, and behavioral traits with risk aversion and insurance uptake, it provides a foundation for dynamic insurance models that reflect real-world behavioral data. Though it does not directly address fatigue, its methodology can be adopted to model fatigue-induced risk. This also includes leveraging large-scale behavioral data to dynamically assess risk. The strengths thus lie in physiological modeling and individualization [38, 39]; some studies focus on operational and behavioral optimization [40, 41]. Lim et al. [42] connect these insights to financial and risk-based decision-making, forming the economic rationale for integrating such systems.

A combined architecture is thus needed. In that, physiological monitoring [38], personal modeling [39], predictive maintenance [40], telematics-based feedback [41], and actuarial modeling [42] are interlinked through edge-based analytics and cloud intelligence. A potential roadmap involves deploying edge AI on vehicles to assess driver fatigue and mechanical stress in real time.

6. AI-powered Intelligent Rest Alert Systems

6.1 Adaptive Alert Mechanisms [43–44]

The exploration of ML models for fatigue prediction, combined with multi-threshold warning systems and integration with vehicle control, represents a critical area in automotive safety and cognitive vehicle technologies. The first key theme, achieving over 98% accuracy in fatigue prediction [2, 6], highlights the remarkable progress made in harnessing ML to interpret complex physiological and behavioral data. Such high-accuracy models indicate that contemporary algorithms can reliably distinguish fatigue states from normal cognitive function. These often utilize multimodal sensor inputs.

The integration of these predictive and warning systems with vehicle control mechanisms pushes the boundary toward true intelligent assistance. EEG-based cognitive architectures [43] that model driver perception, intention, and cognitive state offer a promising method to enhance this integration. EEG signals provide direct insight into the driver's brain activity, capturing nuances of fatigue, distraction, and emotion that are not easily accessible through peripheral sensors alone.

The review of condition-based maintenance (CBM) in rotary machinery [44] shares methods relevant to automotive fatigue detection systems. Both fields rely on advanced sensing, signal processing, and AI-driven prognostics to identify early signs of degradation, whether in machine components or human cognitive states.

Thus, the way forward lies in enhancing the capabilities of high-accuracy ML models, implementing multi-level alerting strategies, and integrating vehicle control systems. This shall be done while utilizing advanced sensing, such as EEG, for rich cognitive insight.

6.2 Personalization and Learning Algorithms

The themes of continuous model refinement based on driver feedback and cross-modal validation through physiological and behavioral data represent essential directions in advancing driver monitoring systems and overall road safety technologies. Continuous refinement, along with real-world driver feedback, ensures that predictive models remain adaptive and personalized, avoiding the discrepancies of static models that fail to generalize well across diverse driver populations and evolving conditions. This iterative feedback mechanism aligns with the findings from the Attention-Driven Reaction Framework (ADRF) [45]. In this, a feedback-driven strategy (RAAO) dynamically optimizes driver attention and reaction predictions. The significant improvements reported include a 25% improvement in reaction time prediction and a 30% increase in decision accuracy.

In continuation of the same, cross-modal validation that integrates physiological signals (such as heart rate, EEG, or eye-tracking) with behavioral indicators (facial expressions, hand movements, or driving patterns) is critical for robust driver state recognition. A review [46] elaborates on the strengths and weaknesses of visual versus non-visual modalities and the increasing role of deep learning approaches in synthesizing these multimodal inputs. The significance of multimodal data fusion extends beyond driver behavior recognition into biomedical and health monitoring contexts [47, 48]. These include early, intermediate, and late fusion. Emerging technologies, such as generative models and large language models [48, 49], promise to reshape multimodal fusion frameworks by enabling richer feature representations and cross-domain knowledge transfer, which can be transformative when applied to real-time driver monitoring systems.

7. Real-World Applications and Case Studies

Akyurek et al. [50] focus on drivers' quality of life. A cross-sectional study involving 62 long-haul drivers identified age, marital status, unemployment, secondary employment, pain, and stress as significant predictors of physical health–related quality of life ($R^2 = 0.663$). This quantified relationship points to the psychosocial burdens drivers face beyond road safety, showing how socioeconomic and occupational factors shape health. The use of instruments like the SF-12 and the Perceived Stress Scale provides a validated view of their mental and physical states. Importantly, the findings recommend occupational therapy interventions and structural reforms, making the study uniquely centered on driver welfare rather than system-wide innovations.

In contrast, another study [51] takes a highly futuristic and system-level approach by examining how metaverse technologies could be integrated into Intelligent Transportation Systems (ITS). Unlike [50], which is empirical and human-focused, this review is conceptual. It proposes that tools like virtual reality, blockchain, digital twins, and AI could revolutionize safety, data management, and testing in ITS. Sterk [52] explores the economic and technological implications of connected car data in the aftermarket sector. It highlights the data access limitations faced by independent players as opposed to OEMs and analyzes different business models. The research focuses on Android Automotive as a promising enabler for third-party innovation.

A further study [53] presents a detailed review of how Electronic Control Units (ECUs) and Inertial Measurement Units (IMUs) function in Advanced Driver Assistance Systems (ADAS). By examining both technological integration and regulatory considerations, and citing real-world cases like Tesla's Autopilot and Volvo's City Safety, the study connects theory with application. The depth of the literature review and inclusion of case studies lend this research credibility. Unlike [33] or [38], which focus on driver behavior and classification, [36] is more aligned with the system architecture and sustainability of automotive innovations.

The validation of low-cost driving simulators [54] adds another layer to the field, focusing on the methodological integrity of human behavior studies. The study combines a literature review with empirical validation comparing real and simulated environments. Despite some incidence of simulator sickness, the simulation was found to reasonably replicate real driving scenarios, affirming its potential for broader behavioral research. While not boasting high classification metrics [55], it addresses a critical concern that emphasizes the credibility of numerous behavioral studies.

Garefalakis et al. [55] offer a quantitative approach to classifying dangerous driving behaviors. It uses both simulated and naturalistic environments and compares four algorithms, namely, SVM, Random Forest (RF), AdaBoost, and MLP neural networks. In simulations, RF achieved an accuracy of 84%, followed by MLP at 82%, whereas in naturalistic settings, accuracy was slightly lower (RF: 75%, AdaBoost: 76.76%). This dual-approach design strengthens the reliability of the findings and provides useful insights for real-world deployment of driving behavior monitoring systems. Unlike [50] or [52], which focus on qualitative or business insights, this study delivers complex performance data with direct application in driver safety interventions.

8. Challenges, Technical Solutions, and Future Work

There are various challenges and solutions in the development, deployment, and safety of autonomous vehicles (AVs), each bringing a distinct perspective to the field [56, 57, 58, 59]. The focus areas range from data privacy and security using blockchain and Federated Reinforcement Learning (FRL), to the role of Explainable Artificial Intelligence (XAI), to the foundational principles of Reliability-Centered Design (RCD), and the exploration of Safety of the Intended Functionality (SOTIF).

Combining the Robot Operating System (ROS), open-source software (OSS), and open-source hardware (OSH) enables a versatile, cost-effective, and collaborative foundation for implementing advanced driver monitoring systems (DMS) in ITS [60–65]. OSS drives spatial analysis and AI traffic forecast and route optimization applications. Based on OSS, OSH refers to physical components with publicly available designs. Many frameworks can handle automobiles, but ROS is the most popular. ROS is an open-source, adaptable library and tool framework for complicated robotics applications. ROS provides the communication layer and module integration structure for ITS and autonomous driving. ROS, OpenCV, dlib, PyTorch, TensorFlow, Autoware, and inexpensive OSH platforms like Raspberry Pi and Arduino are used in open-source ITS driver fatigue and indecision solutions. OSS and OSH in ITS promote collaboration, innovation, and customization in autonomous driving, traffic management, and data analysis. Open Pilot, an open-source autonomous-driving technology, lets compatible vehicles drive themselves. Though less explored than OSS, open design can apply to ITS hardware, including sensors, processing units, and other devices. This method makes ITS more inexpensive and configurable [65].

Alam [66] tackles the increasingly urgent issue of data privacy and cybersecurity in Autonomous Connected Vehicles (ACVs). It highlights the threats posed by cyber-attacks, including remote hacking and sensor data manipulation, which can severely compromise both passenger safety and vehicle functionality. The study proposes a blockchain-enabled Federated Reinforcement Learning (BFRL) framework to secure vast, sensitive datasets generated by ACVs. On the contrary, Kuznietsov et al. [67] focus on XAI as a tool to enhance the safety and trustworthiness of AVs. Unlike [66], which focuses on external threats and infrastructure, this study focuses on internal system transparency. It presents a systematic literature review of data, models, and agency requirements for autonomous-driving AI. The authors introduce a conceptual framework called SafeX that categorizes XAI contributions into five actionable areas, namely, interpretable design, surrogate models, monitoring, auxiliary explanations, and validation. Yazdi [68] considers a comprehensive discussion of RCD and system resilience principles. While not exclusive to autonomous driving, its concepts are directly applicable to AV systems. The study critically considers technical design strategies such as Design of Experiments (DOE), tolerance analysis, and failover mechanisms, making them more prescriptive and engineering-focused.

Wang et al. [69] offer a systematic review of SOTIF, a topic that is becoming increasingly relevant as AVs operate in dynamic, unpredictable environments. Unlike conventional safety that addresses component failure, SOTIF focuses on the functional insufficiencies of correct systems. The paper stands out for its cross-sectional approach, exploring how SOTIF is addressed across development, verification, validation, and operational phases.

Comparatively, the studies vary in their focus. While [66] emphasizes data security frameworks at the network and system level using BFRL, [67] looks at AI systems to promote transparency and trust through XAI. The study in [68] provides foundational engineering practices applicable across sectors but highly relevant to AV reliability, and [69] focuses on the conceptual and practical aspects of functional safety via SOTIF. Thus, each study addresses a different layer of AV development, namely, cybersecurity [66], AI interpretability [67], engineering reliability [68], and systemic safety [69].

Quantitatively, only [66] hints at the scale of data involved in AV operation, highlighting the challenge of processing and securing vast streams of real-time information, though it lacks numerical performance metrics. The other studies are largely qualitative or framework-oriented, though [68] makes use of established methods such as tolerance analysis and DOE, which are inherently data-driven even if not explicitly quantified in the abstract. Critically, each paper responds to a specific gap: [66] bridges cybersecurity and ML through blockchain integration, a novel approach, but still theoretical; [67] addresses the black-box nature of AI systems, offering a structured methodology for interpretability, yet it may face limitations in performance optimization; [68] considers traditional engineering reliability and updates it with modern resilience and redundancy strategies. Its limitation lies in generalization, and [69], by focusing on SOTIF, complements the others by recognizing safety flaws even in technically sound systems, but lacks real-world deployment data.

Looking ahead, there is a need to develop models that can quantify mental workload in real time using continuous physiological data streams such as EEG, ECG, and heart rate variability. These models should prioritize adaptability and responsiveness, thereby suggesting interventions to the individual's current state rather than relying on static thresholds or generalized assumptions [70]. Privacy and data security will also be central concerns as these systems become more integrated into everyday operations. Standardization remains another critical challenge.

Future research should also explore how these fatigue-detection systems can be integrated into vehicle control systems to initiate automated responses such as lane correction or adaptive cruise control. The use of simulation platforms and digital twin technologies could offer a scalable, low-risk way to test these systems across diverse scenarios. Virtual testing environments may also pave the way for more immersive training and intervention tools, potentially extending into metaverse-like applications [71, 72, 73].

9. Conclusion and Future Research Directions

This review highlights the intricate relationship between WBV and the onset of mental fatigue, particularly in high-demand occupations such as driving. Far from being limited to physical discomfort, WBV has been shown to trigger deeper physiological and psychological consequences, including heightened fatigue and burnout. The evidence suggests that the neuromuscular and sensory systems bear significant biological strain even during short-term exposures, which highlights the importance of ergonomic interventions, vehicle condition monitoring, and real-time feedback mechanisms.

The integration of advanced neural network models, especially hybrid CNN-LSTM architectures, has significantly improved the interpretation of temporal and physiological data related to fatigue and performance. These models, when combined with multimodal data from EEG, environmental sensors, and behavioral observations, demonstrate high accuracy in predicting fatigue states. Optimization techniques such as ABC algorithms and edge computing further enhance system responsiveness and scalability. Insights from adjacent fields, such as condition-based maintenance and telematics-based behavioral analytics, further reinforce the potential for intelligent, context-aware systems that can support not just vehicle safety but also holistic driver well-being.

Conflict of Interest

The author(s) declare no potential conflicts of interest with respect to the research, authorship, and/or publication of this article.

Acknowledgments

The authors would like to thank the management, Principal, and Head of Department of Mechanical Engineering at KLS Gogte Institute of Technology (KLS GIT), Belagavi, India, for their constant encouragement and support in carrying out this work. We also wish to acknowledge the learning resources on Ergonomics provided by the Centre for Interdisciplinary Design and Innovation (CIDI) at KLS GIT.

References

[1] Bhuiyan, M.H.U., Fard, M. and Robinson, S.R. (2022). Effects of whole-body vibration on driver drowsiness: A review. *Journal of Safety Research*, *81*, 175–189.

[2] Aguilar, A.J., de la Hoz-Torres, M.L., Martinez-Aires, M.D., Ruiz, D.P., Arezes, P. and Costa, N. (2024). Artificial neural network-based model for assessing the whole-body vibration of vehicle drivers. *Buildings*, *14*(6), 1713.

[3] Estrela, V.V., Deshpande, A., Lopes, R.T., Intorne, A.C., Stutz, D. and Oliveira, L.P. (2023). The building blocks of health 4.0—Internet of Things, big data with cloud and fog computing. In *Intelligent Healthcare Systems*, ed. V. V. Estrela, 24–44. CRC Press.

[4] Estrela, V.V., Laghari, A.A., Lopes, R.T., Khan, A.A., Yin, S. and Deshpande, A. (2023). Introduction to intelligent healthcare in a post-pandemic world. In *Intelligent Healthcare Systems*, ed. V.V. Estrela, 3–23. CRC Press.

[5] Estrela, V.V., Deshpande, A., Lopes, R.T., Yin, S., Andreopoulos, N., Terziev, A. and Laghari, A.A. (2023). On DICOM, HEVC and 3D medical image compression for volumetric theragnostics. In *Intelligent Healthcare Systems*, ed. V.V. Estrela, 203–216. CRC Press.

[6] Estrela, V.V., de Jesus, M.A., Intorne, A.C., Batista, K.K.S., Deshpande, A., Shi, F., et al. (2023). Medical visual theragnostic systems using artificial intelligence (AI)—Principles and perspectives. In *Intelligent Healthcare Systems*, ed. V. V. Estrela, 301–321. CRC Press.

[7] Alsolai, H., Al-Wesabi, F.N., Motwakel, A., et al. Assisting visually impaired people using deep learning-based anomaly detection in pedestrian walkways for intelligent transportation systems on remote sensing images. *Journal of Disability Research*, *2*(2), 49–56.

[8] Nair, A., Patil, V., Nair, R., Shetty, A. and Cherian, M. (2024). A review on recent driver safety systems and its emerging solutions. *International Journal of Computers and Applications*, *46*(3), 137–151.

[9] Estrela, V.V., Grata, E.G.H., Deshpande, A., Sroufe, R., Lopes, R.T., Shi, F., et al. (2023). In-body devices and sensors communication—How implantables, ingestibles, and injectables interact with the Internet. In *Intelligent Healthcare Systems*, ed. V. V. Estrela, 236–258. CRC Press.

[10] Zha, B., Wang, Z., Ma, L., Chen, J., Wang, H., Li, X., Kumar, S. and Min, R. (2024). Intelligent wearable photonic sensing system for remote healthcare monitoring using stretchable elastomer optical fiber. *IEEE Internet of Things Journal*, *11*, 17317–17329.

[11] Kanagamalliga, S., Nath, S.S., Nivetha, T. and Baby Shalini, M. (2024). Enhancing heavy vehicle safety with intelligent driver behaviour monitoring via multi-sensor fusion and

embedded systems. *2024 International Conference on System, Computation, Automation and Networking*, 1–5.

[12] Rahimimoghadam, S., Javan, S., Yari, S., Kheyrkhah, A. and Nourmohamadi, M. (2023). Evaluation of exposure to noise and vibration and its effect on work-related fatigue. *Asian Pacific Journal of Environment and Cancer*, 6(1), 163–168.

[13] Rahimimoghadam, S., Javan, S., Nourmohammadi, M., Bayat, S., Harati, Z., Yari, S. and Kheyrkhah, A. (2023). Evaluation of the relation between noise and vibration exposure with job burnout in city bus drivers. *Asian Pacific Journal of Environment and Cancer*, *6*(1), 15–20.

[14] Abdelrahman, M.M. and Miller, C. (2022). Targeting occupant feedback using digital twins: Adaptive spatial–temporal thermal preference sampling to optimize personal comfort models. *Building and Environment*, *218*, 109090.

[15] Aldhahri, E.A., Almazroi, A.A., Alkinani, M.H. and Alqarni, M.A. (2025). GNN-RMNet: Leveraging graph neural networks and GPS analytics for driver behavior and route optimization in logistics. *PLOS One*, *20*(8), pp. 1–28.

[16] Marinho, C.E.V., Estrela, V.V., Loschi, H.J., Razmjooy, N., Herrmann, A.E., Thiagarajan, Y., Vishnevski, M.P., Monteiro, A.C.B., França, R.P. and Iano, Y. (2019). A model for medical staff idleness minimization. In *Proceedings of the 4th Brazilian Technology Symposium (BTSym'18). BTSym 2018. Smart Innovation, Systems and Technologies, vol 140*, eds Y. Iano, R. Arthur, O. Saotome, V.V. Estrela and H. Loschi, 633–645. Springer, Cham.

[17] Kianpour, P., Gupta, D., Krishnan, K.K. and Gopalakrishnan, B. (2021). Automated job shop scheduling with dynamic processing times and due dates using project management and industry 4.0. *Journal of Industrial and Production Engineering*, *38*(7), 485–498.

[18] Corsini, R.R., Fichera, V., Longo, L. and Oriti, G. (2024). A self-adaptive metaheuristic to minimize the total weighted tardiness for a single-machine scheduling problem with flexible and variable maintenance. *Journal of Industrial and Production Engineering*, *42*(4), 1–18.

[19] Chen, J., Yi, C., Okegbile, S.D., Cai, J. and Shen, X. (2023). Networking architecture and key supporting technologies for human digital twin in personalized healthcare: A comprehensive survey. *IEEE Communications Surveys & Tutorials*, *26*(1), 706–746.

[20] Estrela, V.V., Deshpande, A., Sroufer, R. et al. (2023). Digital twin framework for intelligent healthcare facilities through ISO/IEEE 11073. In *Intelligent Healthcare Systems*, ed. V. V. Estrela, 279–297. CRC Press.

[21] Lyu, X., Azeem Akbar, M. (2025). Driver fatigue warning based on medical physiological signal monitoring for transportation cyber-physical systems. *IEEE Transactions on Intelligent Transportation Systems*, *26*, 14237–14249.

[22] Gupta, B.B., Gaurav, A., Chui, K.T. and Arya, V. (2024). Deep learning model for driver behavior detection in cyber-physical system-based intelligent transport systems. *IEEE Access*, *12*, 62268–62278.

[23] Madhumita, S. and Jeyasekar, A. (2025). Social network analysis in driver monitoring systems: Understanding the impact of predictive models on driver safety. *Social Network Analysis and Mining*, *15*, 34.

[24] Estrela, V.V., Bukhsh, K., Malook Rind, M., Shaikh, S., Khan, A.A., Laghari, A.A., Deshpande, A., Andreopoulos, N. and Terziev, A. (2023). Deep learning as a driving force for better drug development. In *Intelligent Healthcare Systems*, ed. V. V. Estrela, 217–235. CRC Press.

[25] Ahmadzadeh, M., Zahrai, S.M. and Bitaraf, M. (2025). An integrated deep neural network model combining 1D CNN and LSTM for structural health monitoring utilizing multisensor time-series data. *Structural Health Monitoring*, *24*(1), 447–465.

[26] Koşar, E. and Barshan, B. (2023). A new CNN-LSTM architecture for activity recognition employing wearable motion sensor data: Enabling diverse feature extraction. *Engineering Applications of Artificial Intelligence*, *124*, 106529.

[27] Ko, J.E., Kim, S.H., Sul, J.H. and Kim, S.M. (2025). Data reconstruction methods in multi-feature fusion CNN model for enhanced human activity recognition. *Sensors*, *25*(4), 1184.

[28] Deshpande, A., Estrela, V.V. and Razmjooy, N. (2021). *Computational Intelligence Methods for Super-Resolution in Image Processing Applications*. Zurich: Springer Nature.

[29] Estrela, V.V., Razmjooy, N., Monteiro, A.C.B., França, R.P., de Jesus, M.A. and Iano, Y. (2021). A computational intelligence perspective on multimodal image registration for unmanned aerial vehicles (UAVs). In *Metaheuristics and Optimization in Computer and Electrical Engineering. Lecture Notes in Electrical Engineering, vol 696*, eds N. Razmjooy, M. Ashourian and Z. Foroozandeh, 251–274. Springer, Cham.

[30] Sadatdiynov, K., Cui, L., Zhang, L., Huang, J.Z., Salloum, S. and Mahmud, M.S. (2023). A review of optimization methods for computation offloading in edge computing networks. *Digital Communications and Networks*, *9*(2), 450–461.

[31] Babar, M., Khan, M.S., Din, A., Ali, F., Habib, U. and Kwak, K.S. (2021). Intelligent computation offloading for IoT applications in scalable edge computing using artificial bee colony optimization. *Complexity*, 1, 5563531.

[32] Kong, L., Tan, J., Huang, J., Chen, G., Wang, S., Jin, X. and Zeng, P., Khan, M. and Das, S.K. (2022). Edge-computing-driven internet of things: A survey. *ACM Computing Surveys*, *55*(8), 1–41.

[33] Ju, J. and Li, H. (2024). A survey of EEG-based driver state and behavior detection for intelligent vehicles. *IEEE Transactions on Biometrics, Behavior, and Identity Science*. *6*(3), July 2024, pp. 420–434.

[34] Park, S.-H., Han, D.-K., Jang, G.-D. and Lee, S.-W. (2025). Dynamic multi-biosignal fusion for detecting the mental states of drivers and passengers in vehicles. *IEEE Journal of Biomedical and Health Informatics*. *29*(10), Oct 2025, pp. 7104–7115.

[35] Chen, J., Wang, Y., Cui, Y., Wang, H., Polat, K. and Alenezi, F. (2025). EEG-based multi-band functional connectivity using corrected amplitude envelope correlation for identifying unfavorable driving states. *Computer Methods in Biomechanics and Biomedical Engineering*,, 1–13. DOI : https://doi.org/10.1080/10255842.2025.2488502.

[36] Liu, Z., Li, H., Lin, J., Jiao, J., Shen, T., Zhang, B. and Liu, H. (2023). A novel acoustic emission signal segmentation network for bearing fault fingerprint feature extraction under varying speed conditions. *Engineering Applications of Artificial Intelligence*, *126*, 106819.

[37] Ganesh, B.J., Vijayan, P., Vaidehi, V., Murugan, S., Meenakshi, R. and Rajmohan, M. (2024). SVM-based predictive modeling of drowsiness in hospital staff for occupational safety solution via IoT infrastructure. In *Proceedings of the 2nd International Conference on Computer, Communication and Control (IC4)*, 1–5. IEEE.

[38] Ahlström, C., van Leeuwen, W., Krupenia, S., Jansson, H., Finér, S., Anund, A. and Kecklund, G. (2021). Real-time adaptation of driving time and rest periods in automated long-haul trucking: Development of a system based on biomathematical modelling, fatigue and relaxation monitoring. *IEEE Transactions on Intelligent Transportation Systems*, *23*(5), 4758–4766.

[39] Oyebode, O., Fowles, J., Steeves, D. and Orji, R. (2023). Machine learning techniques in adaptive and personalized systems for health and wellness. *International Journal of Human–Computer Interaction*, *39*(9), 1938–1962.

[40] Giordano, D., Giobergia, F., Pastor, E., Macchia, A.L., Cerquitelli, T., Baralis, E., Mellia, M. and Tricarico, D. (2022). Data-driven strategies for predictive maintenance: Lesson learned from an automotive use case. *Computers in Industry*, *134*, 103554.

[41] Gresham, T.R., Kim, J., McDonald, J., Scoggins, N., Mostafavi, M., Park, B.B., Porter, M.D., Duffy, M.E. and Smith, S.A. (2021). Safe and sustainable fleet management with data analytics and training. In *Proceedings of the 2021 Systems and Information Engineering Design Symposium (SIEDS)*, 1–6. IEEE.

[42] Lim, S., Oh, T. and Ngayo, G. (2023). Analyzing factors affecting risk aversion: Case of life insurance data in Korea. *Heliyon*, *9*(10) pp. 1–18.

[43] Mi, P., Yan, L., Cheng, Y., Liu, Y., Wang, J., Shoukat, M.U., Yan, F., Qin, G., Han, P. and Zhai, Y. (2024). Driver cognitive architecture based on EEG signals: A review. *IEEE Sensors Journal*. *24*(22), Nov 2024, pp. 36261–36286.

[44] Kumar, Anil, Gandhi, C.P., Tang, H., Sun, W. and Xiang, J. (2023). Latest innovations in the field of condition-based maintenance of rotatory machinery: A review. *Measurement Science and Technology*, *35*(2), 022003.

[45] Xianyou, Y. and Bai, J. (2025). The impact of physical exercise on driver attention reaction time and decision making. *IEEE Access*. Vol 13, pp. 109867–109882.

[46] Tan, D., Tian, W., Wang, C., Chen, L. and Xiong, L. (2024). Driver distraction behavior recognition for autonomous driving: Approaches, datasets and challenges. *IEEE Transactions on Intelligent Vehicles*. *9*(12), Dec 2024, pp. 8000–8026.

[47] Duan, J., Xiong, J., Li, Y. and Ding, W. (2024). Deep learning based multimodal biomedical data fusion: An overview and comparative review. *Information Fusion*, Vol. 112, Dec 2024, 102536.

[48] Melnychuk, V. and Feuerriegel, S. (2025). GDR-learners: Orthogonal learning of generative models for potential outcomes. ArXiv, abs/2509.22953.

[49] Kulkarni, D.S., Deshpande, A.B., Bhandurge, S.A. and Estrela, V.V. (2025). Impact of textual length on lexicon-based sentiment classification for Hindi reviews. *Cureus Journal of Computer Science*, *2*, es44389-025-05622-w.

[50] Akyurek, G., Ozturk, L.K. and Gurlek, S. (2025). Investigation of factors predicting quality of life of drivers working on long-haul transport: Pain, fatigue, stress and work role functions. *International Journal of Occupational Safety and Ergonomics*, 1–7. DOI : https://doi.org/10.1080/10803548.2025.2474835

[51] Sarwatt, D.S., Lin, Y., Ding, J., Sun, Y. and Ning, H. (2024). Metaverse for intelligent transportation systems (ITS): A comprehensive review of technologies, applications, implications, challenges and future directions. *IEEE Transactions on Intelligent Transportation Systems*, *25*(7), 6290–6308.

[52] Sterk, F. (2024). Leveraging car connectivity in the automotive aftermarket and beyond. In *Automotive Aftermarket: Global and Interdisciplinary Perspectives*, Maria Elo, Fotios Katsardis. 357–370. Cham: Springer International Publishing.

[53] Tadimarri, A., Katari, M., Gowrisankar, K. and Shanmugam, L. (2024). Driving towards safety: the role of ECUs and IMUs in advanced driver-assistance systems. *International Journal for Multidisciplinary Research*, *6*(2), 1–14.

[54] Andriola, C., Di Rado, G.R., García, D.S.P. and Nodari, C.T. (2025). Classification of driving simulators validation: A case study using an immersive driving simulator. *Accident Analysis & Prevention*, *213*, 107944.

[55] Garefalakis, T., Michelaraki, E., Roussou, S., Katrakazas, C., Brijs, T. and Yannis, G. (2024). Predicting risky driving behavior with classification algorithms: Results from a large-scale field-trial and simulator experiment. *European Transport Research Review*, *16*(1), 65.

[56] He, Z., Xu, R., Wang, B., Meng, Q., Tang, Q., Shen, L., Tian, Z. and Duan, J. (2025). Integrated blockchain and federated learning for robust security in internet of vehicles networks. *Symmetry*, *17*(7), 1168.

[57] Estrela, V.V., de Jesus, M.A., Intorne, A. C., Batista, K. K. S., Deshpande, A., Shi, F., Khan, A. A. and Oliveira, L. P. (2023). Blockchain technology enabling better services in the healthcare domain. In *Intelligent Healthcare Systems*, ed. V. V. Estrela, 135–158. CRC Press.

[58] Gaur, L. and Sahoo, B.M. (2022). Introduction to explainable AI and intelligent transportation. In *Explainable Artificial Intelligence for Intelligent Transportation Systems*. Loveleen Gaur, Biswa Mohan Sahoo. pp. 1–25. Springer, Cham.

[59] Weerasinghe, N., Usman, M.A., Hewage, C., Pfluegel, E. and Politis, C. (2023). Threshold cryptography-based secure vehicle-to-everything (V2X) communication in 5G-enabled intelligent transportation systems. *Future Internet*, *15*(5), 157.

[60] Alvarez Oviedo, A., Mamani Villanueva, J.F., Echaiz Espinoza, G.A., Villanueva, J.M.M., Salazar, A.O. and Villarreal, E.R.L. (2025). Design of a system for driver drowsiness detection and seat belt monitoring using Raspberry Pi 4 and Arduino Nano. *Designs*, *9*(1), 11.

[61] Bertoli, G.C., Saotome, O. and Estrela, V.V. (2020). Computer vision in UAV using ROS. In *Imaging and Sensing for Unmanned Aircraft Systems*, Vol. 1, 9, eds V.V. Estrela, J. Hemanth, O. Saotome, G. Nikolakopoulos and R. Sabatini, 217–260. London: IET.

[62] Testouri, M., Elghazaly, G. and Frank, R. (2024). RoboCar: A rapidly deployable open source platform for autonomous driving research. *IEEE Intelligent Transportation Systems Magazine*, *17*, 83–95.

[63] Burdziakowski, P., Razmjooy, N., Estrela, V.V. and Hemanth, J. (2020). Open source software (OSS) and hardware (OSH) in UAVs. In *Imaging and Sensing for Unmanned Aircraft Systems*, Vol. 2, 3, eds V. V. Estrela, J. Hemanth, O. Saotome, G. Nikolakopoulos and R. Sabatini, 49–66. London: IET.

[64] Visconti, P., Rausa, G., Del-Valle-Soto, C., Velázquez, R., Cafagna, D. and De Fazio, R. (2025). Innovative driver monitoring systems and on-board-vehicle devices in a smart-road scenario based on the Internet of Vehicle paradigm: A literature and commercial solutions overview. *Sensors*, *25*(2), 562.

[65] Li, Z., Seering, W., Yang, M.C. and Eesley, C.E. (2021). Understanding the motivations for open-source hardware entrepreneurship. *Design Science*, *7*, e19.

[66] Alam, T. (2024). Data privacy and security in autonomous connected vehicles in smart city environment. *Big Data and Cognitive Computing*, *8*(9), 95.

[67] Kuznietsov, A., Gyevnar, B., Wang, C., Peters, S. and Albrecht, S.V. (2024). Explainable AI for safe and trustworthy autonomous driving: A systematic review. *IEEE Transactions on Intelligent Transportation Systems*. *25*(12), Dec 2024, pp. 19342–19364.

[68] Yazdi, M. (2024). Reliability-centered design and system resilience. In *Advances in Computational Mathematics for Industrial System Reliability and Maintainability*, Hoang Pham 79–103. Cham: Springer Nature Switzerland.

[69] Wang, H., Shao, W., Sun, C., Yang, K., Cao, D. and Li, J. (2024). A survey on an emerging safety challenge for autonomous vehicles: Safety of the intended functionality. *Engineering*, *33*, 17–34.

[70] Nair, A., Patil, V., Nair, R., Shetty, A. and Cherian, M. (2024). A review on recent driver safety systems and its emerging solutions. *International Journal of Computers and Applications*, *46*, 137–151.

[71] Deshpande, A., Estrela, V.V., Patavardhan, P. and Kallimani, G. (2023). Super-resolution in a world of scarce resources for medical imaging applications. In *Edge-AI in Healthcare: Trends and Future Perspectives*, eds S. Vyas et al.. CRC Press.

[72] Llagostera-Brugarola, E., Corpas-Marco, E., Victorio-Vergel, C., Lopez-Aguilera, E., Vázquez-Gallego, F. and Alonso-Zarate, J. (2025). A digital twin for intelligent transportation systems in interurban scenarios. *Applied Sciences*, *15*(13), 7454.

[73] Antoun, M., Elhajj, I.H., Sayour, M. et al. (2025). Interactive digital twins enabling responsible extended reality applications. *Scientific Reports*, *15*, 34539.

10 AR and AI for Driver Assistance and Navigation

Sakshi Hatti[1*], Prathmesh Redekar[2], Aditya Patil[3], Tanishq Jain[4], Yash Mulay[5] and Prakash Sonwalkar[6]

The combination of augmented reality (AR) and artificial intelligence (AI) is changing driver assistance and navigation systems. This shift leads to safer, more efficient, and intuitive driving experiences. This chapter reviews how AR and AI are currently used in smart vehicle technologies. It focuses on practical examples, system design, and innovations. AR tools, such as head-up displays (HUDs) and windshield projections, give real-time visual guidance by placing relevant information in the driver's line of sight. At the same time, AI helps with object detection, behavior analysis, predictive decision-making, and flexible route planning. They improve situational awareness and support better decision-making on the road. The chapter discusses the technical framework underlying these systems, including sensor fusion, spatial computing, and machine learning models. This review offers insights for researchers, developers, and policymakers. It encourages the development of user-focused, intelligent navigation systems and highlights the need for cooperation to enhance smart mobility.

1. Introduction

In recent years, the use of the automobile as the primary mode of transportation has increased, and driving has become an important part of daily life. Every accident is one too many, especially when it results in injury or even loss of life. Given that the global traffic volume has increased to around 1 billion vehicles, road safety has become a key challenge for society, industry, and politicians. Nevertheless, navigation can be perceived as a broader ecosystem encompassing various types of vehicles and their interactions with one another, with other road users, infrastructure, and the environment.

An integrated intelligent transportation system (ITS) unites earth, maritime, aerial, and space domains with augmented reality (AR), merging real-time transport

[1] Department of Computer Science and Engineering, Jain College of Engineering and Research, Belagavi, Karnataka, India.

[2,3,6] Department of Computer Science and Engineering (AIML), Jain College of Engineering and Research, Affiliated to Visvesvaraya Technological University (VTU), Belagavi, Karnataka, India.

[4-5] Department of Computer Science and Engineering (AIML), KIT's College of Engineering, Kolhapur, Maharashtra, India.

* Corresponding author: hattisakshi@gmail.com

DOI: 10.1201/9781003648147-10

data with users' physical environments. Powered by artificial intelligence (AI) and sensor fusion, it manages traffic, enhances safety by visualizing hazards, and provides real-time navigational guidance. Applications span air traffic control, maritime navigation, and space exploration, using AR + AI to display critical details such as routes and hazards. Use cases include [1–10]

i. Ground transportation is the most widespread arena for vehicular research with AR+AI.
ii. In maritime navigation, AR can project a navigable route directly onto the water's surface, display water depth information, and highlight other vessels or obstacles, even in low-visibility conditions.
iii. The aviation department can monitor traffic, while AR + AI can provide pilots and operators with real-time data on other aircraft, flight paths, and potential conflicts.

Current limitations and future developments entail aspects like sensor fusion and localization accuracy, data interoperability, and standardization calls for a truly integrated system, standardized protocols, and frameworks for data exchange and AR visualization.

Advancements in AI and AR are changing driver assistance and navigation. AI helps with quick decision-making [11–13]. It analyzes complex traffic situations by spotting obstacles, predicting actions, and adjusting routes on the fly.

At the same time, AR is being added to vehicle dashboards to increase awareness. Important information, such as speed, step-by-step directions, and collision alerts, is displayed within the driver's view, as shown in Figure 10.1. Technologies like head-up displays (HUDs) and AR-enhanced windshields give clear, real-time guidance while reducing distractions. These solutions address the problems of traditional navigation systems [14–18], which can hide what is happening on the road and lead to bad driving decisions.

Fig. 10.1 Demo application of AI and AR in decision-making

Even though technology has come a long way, there are still a lot of problems with current navigation and assistance systems. Drivers often have too much information to process and are easily distracted [15], especially in complex or new situations, which lead to bad choices. Also, there is a gap between digital navigation data and the real world, which makes it harder for drivers to process information. This problem is made worse by the lack of smooth AR integration. These limits are very dangerous for road safety because of human mistakes, tiredness, and bad judgment. So, the combination of AR and AI is a good idea right now. It combines smart data processing with immersive visualization to make driving experiences intuitive, safe, and efficient [19–22]. Vehicle operator distraction, insecurity, and inattention are significant causes of vehicle accidents that can be addressed with AI and AR. Metaverse is widely employed in other industries, but its transportation potential remains untapped [23–27]. The cyber-physical universe can be connected to ITS applications such as AR, digital twins [28], blockchain [29–31], AI, and new human-machine interfaces [33], among others.

This chapter discusses how the combined effect of AR and AI can significantly improve road safety and streamline traffic flow. AR-based navigation keeps drivers focused ahead while minimizing distractions, while AI models anticipate traffic congestion and suggest alternate routes in real time.

2. Methodology

2.1 Data Collection Framework

To establish a multidimensional data-collection model, the proposed research methodology will use a combination of qualitative and quantitative data to develop an inclusive AR- and AI-based driver assistance and navigation technology system.

2.2 Input Sensors and Devices

It is based on the usage of high-end algorithms and machine learning activities to process the data flows of real-time sensors and input devices, i.e., LiDAR, radar, GPS, accelerometers, and cameras, and the in-car AR-HUD informing drivers about the most significant details, e.g., navigation guidance, traffic situation results, updates, and on-time warnings, thereby potentially improving the overall driving experience and minimizing the risk of accidents [34]. The GPS will provide location and mapping information for the navigational area, allowing the creation of much more accurate navigation pathways and predictive traffic models [35]. The combination of the input devices and the sensors will provide a robust, stable information-gathering system capable of developing advanced driver-assist and guidance applications. In-car AR displays and HUDs will update drivers in real time with direction information, traffic conditions, and potential hazards, improving their driving experience and safety [35]. Figure 10.2 shows the distribution of functional categories in an advanced driver assistance system using AI, sensors, and real-time interfaces.

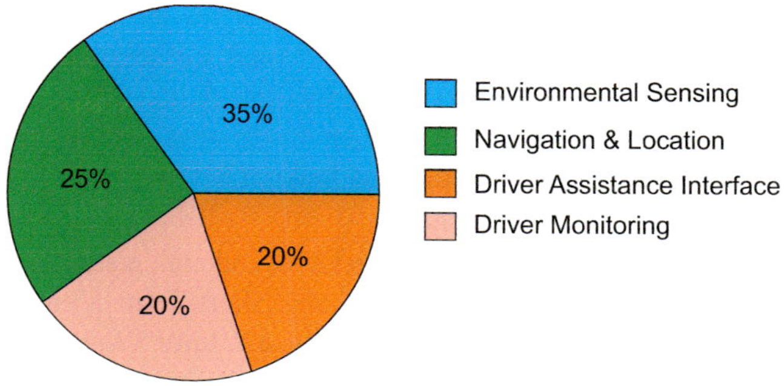

Fig. 10.2 Functional distribution of technologies in intelligent driver systems

2.3 Data Types

The information that can be obtained from sensors or the kinds of information they can provide has several limitations [36, 37]. Health surveillance involves psychological, physiological, and neurological tests that can be used to evaluate various abnormalities in driver behavior [38, 39]. Cameras constitute a major advantage over other sensor types since they can distinguish among colors and/or textures. Also, several cameras at various locations around the vehicle provide a 360° view of the surroundings, offering a bird's-eye view of the vehicle and enabling it to recognize local features such as surrounding cars, people, roads, road markings, and signposts, among others [41].

2.4 Preprocessing

Raw sensor data are contaminated by noise from environmental factors such as road vibrations, lighting, and weather changes. As a precaution, filtering methods such as Kalman filters for visual input are applied [42]. Those techniques reduce temporary errors and improve control detection. The multi-sensor configurations address decimation and sampling delays. Therefore, sensor information may be precisely synchronized in real time using timestamp-based synchronization systems. Sensor fusion algorithms, such as Extended Kalman Filters and probabilistic fusion models, fuse camera, LiDAR, and GPS sensors to generate a composite representation of the surroundings of the vehicle [43, 44], making AI-based interpretation more reliable and better aligning with the physical world through an AR overlay.

2.5 Feature Engineering

(i) *Pulling of AR Overlays:* Lane boundaries, traffic signs, and points of interest (POI) are the key visual cues retrieved using convolutional neural networks (CNNs) trained on datasets such as Cityscapes [45]. Such overlays are incorporated into AR interfaces to provide intuitive visual guidance to drivers.

(ii) *Road Scene Parsing:* Semantic segmentation divides an input image into categories, including car, pedestrian, and road surface, with class labels. This task is performed using state-of-the-art architectures, DeepLabV3+ and U-Net, which are trained on autonomous-driving datasets [46], enabling context-aware AI decision-making and making it easier to place meaningful AR content.

In proactive assistance, it is essential to know and anticipate driver behavior. Historical movement data are applied to Long Short-Term Memory (LSTM) networks and other sequence models to forecast future routes and driving intentions [47]. These forecasts can be used for real-time navigation decisions and for forward-looking AR views, e.g., turn signals or hazard warnings. Figure

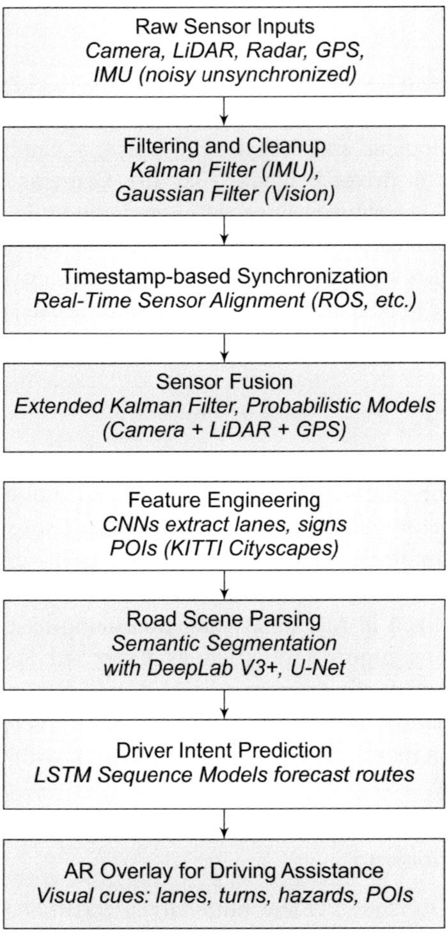

Fig. 10.3 End-to-end data processing pipeline for AR and AI-based driving assistance systems

10.3 shows the end-to-end data processing pipeline for AR and AI-based driving assistance systems.

2.6 AI Model

The AI models may consist of one of the following or combinations of all:

- *CNNs*: They are a narrower form of neural networks designed to process structural data and for visual recognition problems. This network is influenced by the human visual system, which uses stacks of interconnected neurons to detect and analyze features hierarchically. CNNs have been considered the primary instrument in interpreting data [48].
- *Recurrent Neural Networks (RNNs) and LSTM*: RNNs and LSTMs are great at predicting the future trajectories of other cars on the street by leveraging past information. By using driving data to train RNNs, they can simulate driver behavior, enabling autonomous vehicles to act similarly to human drivers. RNNs and LSTMs help combine data from various sensors and report the vehicle's position by fusing sensor signals [48].
- *Adaptive Decision Making via Reinforcement Learning*: The main advantage of reinforcement learning (RL) is that it can adapt in real time to driver assistance systems, where decisions are agent-decoded, and can learn a potentially optimal control policy by interacting with dynamic environments. In contrast to other supervised models, RL can learn temporal and long-term rewards, and therefore can be applied to more complex tasks in self-driving cars, such as lane merging, adaptive speed control, and collision avoidance [49].

2.7 AR Techniques

The embodiment of AR in driver assistance programs is based on accurate geospatial representation and semantic visual annotations to support the driver by raising awareness. Real-time object labeling provides semantic details about adjacent objects, which are overlaid on the driver's view to enable recognition and response, assisting drivers more promptly. These overlays are created from fused camera, LiDAR, GPS, Inertial Navigation System, Global Navigation Satellite System, and Inertial Measurement Unit data [50–54] to ensure proper positioning and tracking.

3. Conclusion

The driving experience is changing as AR and AI are combined in contemporary driver assistance and navigation systems. Furthermore, AR would build a seamless, real-time network that uses digital overlays to deliver information, improve safety, and increase efficiency across all modes of transport, i.e., terrestrial, maritime, aerial, and spatial. AR presents real-time, valid data, such as routes, speed limits, and danger alerts, directly in front of vehicle users, keeping their eyes on the road. Meanwhile, AI remains in the background, constantly analyzing the environment to provide real-time, smart support. In combination, they can enable drivers to act faster, feel less mentally stressed, and drive more safely by helping to overcome driver distraction and fatigue. The impact is not limited to individual vehicle users. AR also makes the self-driving options less learnable and credible to ordinary drivers. Integrating AR and AI can be a savvy move, given the increased traffic and the need to establish a pattern of road safety going forward.

References

[1] Brata, K.C., Funabiki, N. and Y.Y.F. Panduman, (2024). An enhancement of outdoor location-based augmented reality anchor precision through VSLAM and Google Street View. *Sensors*, *24*(4), 1161.

[2] Singh, J., Tyagi, N., Singh, S. and A.A. Al Zubi (2023). Enhancing indoor navigation in intelligent transportation systems with 3D RIF and Quantum GIS. *Sustainability*, *15*(22), 15833.

[3] Razmjooy, N., Estrela, V.V. and Sabatini R. (2020). Vision in micro-aerial vehicles. In *Imaging and Sensing for Unmanned Aircraft Systems*, Vol. 1, eds V.V. Estrela, J. Hemanth, O. Saotome, G. Nikolakopoulos and R. Sabatini, 173–216. London: IET.

[4] Afrin, T., Yodo, N., Dey, A. and Aragon, L.G. (2024). Advancements in UAV-enabled intelligent transportation systems: A three-layered framework and future directions. *Applied Sciences*, *14*(20), 9455.

[5] Zhang, R., Ai, Y., Li, S., Hu, J., Hao, J. and Pan, M. (2025). Application of augmented reality in waterway traffic management using sparse spatiotemporal data. *Applied Sciences*, *15*(4), 1710.

[6] Estrela, V.V., Deshpande, A. and Stutz, D. (2023). Ad hoc networks in healthcare intelligent transportation systems humanitary relief—MANETs, VANETs, and FANETs. In *Intelligent Healthcare Systems*, ed. V. V. Estrela, 66–89. CRC Press.

[7] Bellemare, M.G., Candido, S., Castro, P.S., Gong, J., Machado, M.C., Moitra, S., Ponda, S.S. and Wang, Z. (2020). Autonomous navigation of stratospheric balloons using reinforcement learning. *Nature*, *588*, 77–82.

[8] Greaves, J.A. and Scheeres, D.J. (2024). Spacecraft to spacecraft absolute tracking for autonomous navigation of a distributed space system from relative sensors. *The Journal of the Astronautical Sciences* (2024) 71: 46. https://doi.org/10.1007/s40295-024-00463-6

[9] Ma, Y., Xue, R., Qiang, X. and Gong, Y. (2025). Cooperative localization for multi-airship navigation network with hybrid kernel particle filter. *Chinese Journal of Aeronautics*. https://doi.org/10.1016/j.cja.2025.103843

[10] Jumani, A.K., Laghari, A.A. and Rashid A. Laghari, (2025). Image processing-based via agridrones for agriculture fields. In *Intelligent Designs, Innovations and Sustainability in Agriculture 4.0*, eds A. Deshpande, V. V. Estrela and A. Khan, 214–241. Boca Raton: CRC Press.

[11] Parvatha, Nagaraj. (2024). AI-Augmented Reality: Bridging the Gap between Virtual and Physical Worlds in Real-Time. *International Journal of Scientific Research and Management* (IJSRM). 12. 1362–1373. 10.18535/ijsrm/v12i07.ec08.

[12] Cao, M., Zheng, L., Jia, W. and Liu, X. (2021). Joint 3D reconstruction and object tracking for traffic video analysis under IoV environment. *IEEE Transactions on Intelligent Transportation Systems*, *22*, 3577–3591.

[13] Estrela, V.V., Khan, A.A. and Aftab A. Shaikh (2023). Some issues regarding content-based image retrieval (CBIR) for remote healthcare theradiagnosis. In *Intelligent Healthcare Systems*, ed. V.V. Estrela, 110–134. Boca Raton: CRC Press.

[14] Butt, M. and Shafique, M. (2025). A literature review: AI models for road safety for prediction of crash frequency and severity. *Discover Civil Engineering, 2*, 99.

[15] Brown, J.E. and Davis, C.L. (2022). Ethical considerations in driver monitoring using AI technologies. *IEEE Transactions on Technology and Society, 3*(2), 112–125.

[16] Vania Vieira Estrela & Maria Aparecida de Jesus & Jenice Aroma & Kumudha Raimond & Sandro R. Fernandes & Nikolaos Andreopoulos & Edwiges G.H. Grata & Andrey Terziev & Ricardo Tadeu Lopes & Anand Desh, 2021. "Motion Estimation Role in the Context of 3D Video," *International Journal of Multimedia Data Engineering and Management* (IJMDEM), IGI Global Scientific Publishing, vol. *12*(3), pp. 16–38, July.

[17] Estrela, V.V., Saotome, O., Loschi, H.J., Hemanth, D.J., Farfan, W.S., Aroma, R.J., Saravanan, C. and Grata, E.G.H. (2018). Emergency response cyber-physical framework for landslide avoidance with sustainable electronics. *Technologies, 6*, 42.

[18] Zhang, S., Wang, W., Li, H. and Zhang, S. (2023). E-detector: Asynchronous spatio-temporal for event-based object detection in intelligent transportation system. *ACM Transactions on Multimedia Computing, Communications and Applications, 20*, 1–20.

[19] Wan, Z., Zou, J.M. and Liu, S. (2024). Detection algorithm for multimodal data fusion in smart transportation. 2024 6th International Academic Exchange Conference on Science and Technology Innovation, 2090–2094.

[20] Ibset, A., Bencharif, L., Liu, H., Qi, W., Su, H. and Alfayad, S. (2025). Lower-limb full-body motion acquisition using wearable multisensor integration for human-robot interaction. 2025 Eighth International Conference on Advances in Biomedical Engineering, 1–4.

[21] Estrela, V.V., Razmjooy, N., Monteiro, A.C.B., França, R.P., de Jesus, M.A. and Iano, Y. (2021). A computational intelligence perspective on multimodal image registration for unmanned aerial vehicles (UAVs). In *Metaheuristics and Optimization in Computer and Electrical Engineering*, eds N. Razmjooy, M. Ashourian and Z. Foroozandeh. Lecture Notes in Electrical Engineering, Vol. 696. Springer, Cham.

[22] Feng, S., Li, X., Yan, Z., Xia, C., Li, S., Wang, X. and Zhou, Y. (2025). Tightly coupled integration of LiDAR and Vision for 3D multiobject tracking. *IEEE Transactions on Intelligent Vehicles, 10*, 399–412.

[23] Obayya, M.I., Al-Wesabi, F.N., Alabdan, R., Khalid, M., Assiri, M., Alsaid, M.I., Osman, A.E. and Alneil, A.A. (2024). Artificial intelligence for traffic prediction and estimation in intelligent cyber-physical transportation systems. *IEEE Transactions on Consumer Electronics, 70*, 1706–1715.

[24] Sarwatt, D.S., Lin, Y., Ding, J., Sun, Y. and Ning, H. (2024). Metaverse for intelligent transportation systems (ITS): A comprehensive review of technologies, applications, implications, challenges and future directions. *IEEE Transactions on Intelligent Transportation Systems*, 25, 6290–6308.

[25] Laghari, A.A. et al. (2024). Quality of experience assessment in virtual/augmented reality serious games for healthcare: A systematic literature review. *Technology and Disability, 36*, 17–28.

[26] Dahri, F.H., Laghari, A.A., Dahri, N.A., Abro, G., Estrela, V.V., de Oliveira, G.G., Iano, Y. and Chuma, E.L. (2024). Real-time detection of UAV small target's infrared imaging via OSTD-YOLOv8. 2024 IEEE URUCON, 1–5.

[27] Wang, J., Chen, Y., Ji, X., Dong, Z., Gao, M. and Sing Lai, C. (2024). Metaverse meets intelligent transportation system: An efficient and instructional visual perception framework. *IEEE Transactions on Intelligent Transportation Systems, 25*, 14986–15001.

[28] Estrela, V.V., Deshpande, A., Sroufe, R., Lopes, R.T., Grata, E.G.H., Andreopoulos, N., Terziev, A. and Laghari, A.A. (2023). Digital twin framework for intelligent healthcare facilities through ISO/IEEE 11073. In *Intelligent Healthcare Systems*, ed. V. V. Estrela, 279–297. Boca Raton: CRC Press.

[29] Khan, A.A., Rashid, A. Laghari et al. (2025). Quantum computing empowering blockchain technology with post quantum resistant cryptography for multimedia data privacy preservation in cloud-enabled public auditing platforms. *Journal of Cloud Computing*, *14*, 43.

[30] Bukhsh, K., Rind, M.M., Shaikh, S., Khan, A.A., Laghari, A.A., Estrela, V.V. and Andreopoulos, N. (2025). Blockchain and Internet of Things (IoT) enabled secure communication protocols for smart agriculture. In *Intelligent Designs, Innovations and Sustainability in Agriculture 4.0*, eds A. Deshpande, V.V. Estrela and A. Khan, 267–284. Boca Raton: CRC Press.

[31] Khuwuthyakorn, P., Lakhan, A., Majumdar, A. and Thinnukool, O. (2025). Blockchain-enabled self-autonomous intelligent transport system for drone task workflow in edge cloud networks. *Algorithms*, *18*(8), 530.

[32] Reddy, V.R., Krishna, P.V., Anil, C., Swathi, A., Kalaiselvan, S. and Chiranjeevi, S. (2025). Fostering Human Trust in Intelligent Cyber-Physical Environments. (2025). *International Journal of Environmental Sciences*, 8378–8383. https://doi.org/10.64252/2j707j20

[33] Singh, A. (2025). Enhancing autonomous vehicle safety with augmented reality. AZoRobotics, 2024. Available at https://www.azorobotics.com/Article.aspx?ArticleID=693 (accessed on 26 June 2025).

[34] Tashko, P., Kjosevski Rizov, M.P. and Tashevski, R.P. (2017). Driver assistance systems in vehicles using augmented reality—benefits and challenges. *Trans Motauto World*, *2*(5), 201–206.

[35] Al Smadi, T. and Al-Maitah, M. (2020). Artificial intelligence technology for safe driver assistance system. *International Journal of Computer Aided Engineering and Technology*, *13*(1–2): 183–191..

[36] Elamrani Abou Elassad, Z., Mousannif, H., Al Moatassime, H. and Karkouch, A. (2020). The application of machine learning techniques for driving behavior analysis: A conceptual framework and a systematic literature review. *Engineering Applications of Artificial Intelligence*, 87,103312.

[37] Yan, S., Teng, Y., Smith, J. and Zhang, B. (2016). Driver behavior recognition based on deep convolutional neural networks. In *Proceedings of the 12th International Conference on Natural Computation, Fuzzy Systems and Knowledge Discovery*, 636–641.

[38] Balico, L., Loureiro, A., Nakamura, E., Barreto, R., Pazzi, R. and Oliveira, H. (2018). Localization prediction in vehicular ad hoc networks. *IEEE Communications Surveys & Tutorials*, *20*(4), 2784–2803.

[39] Kaye, S., Lewis, I. and Freeman, J. (2018). Comparison of self-report and objective measures of driving behavior and road safety: A systematic review. *Journal of Safety Research*, *65*, 141–151.

[40] Shahverdy, M., Fathy, M., Berangi, R. and Sabokrou, M. (2020). Driver behavior detection and classification using deep convolutional neural networks. *Expert Systems with Applications*, *149*, 113240.

[41] Baccari, S., Touati, H., Hadded, M., Ghazzai, H. and Elhade, M. (2024). Anomaly detection in connected and autonomous vehicles: A survey, analysis, and research challenges. *IEEE Access*, *12*.

[42] Welch, G. and Bishop, G. (2006). An introduction to the Kalman filter. Technical Report, Department of Computer Science, University of North Carolina at Chapel Hill.

[43] Paden, B., C'ap, M., Yong, S., Yershov, D. and Frazzoli, E. (2016). A survey of motion planning and control techniques for self-driving urban vehicles. *IEEE Transactions on Intelligent Vehicles*, *1*(1), 33–55.

[44] Durrant-Whyte, H. and Bailey, T. (2006). Simultaneous localization and mapping: Part I. *IEEE Robotics & Automation Magazine*, *13*(2), 99.

[45] Geiger, A., Lenz, P. and Urtasun, R. (2012). Are we ready for autonomous driving? The KITTI vision benchmark suite. *IEEE Conference on Computer Vision and Pattern Recognition*, Providence, RI, 3354–3361.

[46] Chen, Liang-Chieh & Zhu, Yukun & Papandreou, George & Schroff, Florian & Adam, Hartwig. (2018). Encoder-Decoder with Atrous Separable Convolution for Semantic Image Segmentation: 15th European Conference, Munich, Germany, September 8–14, 2018, Proceedings, Part VII. 10.1007/978-3-030-01234-2_49.
[47] Deo, N. and Trivedi, M. (2018). Convolutional social pooling for vehicle trajectory prediction. *IEEE Conference on Computer Vision and Pattern Recognition Workshops (CVPRW)*, Salt Lake City, UT, 1549.
[48] Hephzibah, J. andMadhumitha, S. (2023). Enhancing safety, navigation, and user experience with AR integration in 5G-powered autonomous vehicles. *International Journal of Engineering, Management and Humanities*, *4*(5), 428–435.
[49] Lillicrap, T., Hunt, J., Pritzel, A. and Wierstra, D. (2015). Continuous control with deep reinforcement learning. arXiv preprint arXiv:1509.02971.
[50] Zhang, Y., Chu, L. and Y. Mao, (2024). A vision/inertial navigation/global navigation satellite integrated system for relative and absolute localization in land vehicles. *Sensors*, *24*(10). DOI:10.3390/s24103079
[51] Meng, X., Wang, H. and Liu, B. (2017). A robust vehicle localization approach based on GNSS/IMU/DMI/LiDAR sensor fusion for autonomous vehicles. *Sensors*, *17*, 2140.
[52] Estrela, V.V., Deshpande, A., Stutz, D., de Assis, J.T., Laghari, A.A., Shi, F., Lin, Yu-Da. (2023). 6G in healthcare—Anticipating needs and requirements. In *Intelligent Healthcare Systems*, ed.. V.V. Estrela, 159–181. Boca Raton: CRC Press.
[53] Estrela, V.V., Hemanth, J., Saotome, O., Nikolakopoulos, G. and Sabatini, R. (2020). *Imaging and sensing for unmanned aircraft systems: Control and performance*, Vol. 1. London: IET.
[54] Fuentes-Pacheco, J., Ruiz-Ascencio, J., Rendon-Mancha, J. (2015). Visual simultaneous localization and mapping: A survey. *Artificial Intelligence Review*, *43*(1), 55–81.

11 The Invisible Highway: Integrating Artificial Intelligence with 6G Next-Generation Mobility

Anjanabhargavi Kulkarni[1*], R. H. Goudar[2] and
Gautam Arjun Dematti[3]

This chapter explores the integration of artificial intelligence (AI) with next-generation wireless networks (5G and 6G) to revolutionize global transportation systems. It investigates how AI-enabled autonomous systems, powered by 6G's terahertz bandwidth and ultra-low latency, improve traffic efficiency, reduce human error, and support vehicle-to-everything (V2X) communication. Key discussions include ethical AI dilemmas, infrastructure sustainability, and variations in global policy. This chapter interrogates the symbiotic convergence of AI and sixth-generation (6G) wireless networks as the foundational paradigm for next-generation mobility ecosystems: urbanization pressures, safety imperatives, and climate urgency demand transformative solutions beyond legacy infrastructure. We demonstrate how AI's predictive intelligence—powered by 6G's terahertz-band connectivity (0.1–10 THz), submillisecond latency, and holographic beamforming (HB)—enables real-time, self-optimizing transportation networks. Autonomous vehicles (AVs) leverage this association through sensor fusion, V2X communication, and neuromorphic computing (e.g., Tesla's Dojo chip, which reduces training energy by 50%), achieving 89% reductions in accidents in trials. With international case studies, tables, charts, and infographics, the chapter offers a future-forward vision where AI, quantum computing, and self-optimizing networks form the foundation of inclusive, resilient, and carbon-neutral mobility ecosystems.

1. Introduction: The AI–Network Convergence

The dawn of the twenty-first century has ushered in an era where artificial intelligence (AI) and next-generation networks are no longer isolated technological marvels but intertwined forces driving the future of mobility. This convergence represents a change in thinking, transforming how humans and machines navigate physical and digital landscapes. At its core, AI–network constructive collaboration enables real-time data processing, predictive analytics, and decentralized decision-

[1-2] Department of Computer Science and Engineering, Visvesvaraya Technological University, Belagavi, Karnataka, India.

[3] Department of Computer Science and Engineering, Angadi Institute of Technology and Management, Belagavi, Karnataka, India

* Corresponding author: anjanapramod2022@gmail.com

DOI: 10.1201/9781003648147-11

making-capabilities essential for autonomous vehicles (AVs), smart infrastructure, and seamless urban mobility. As 5G matures and 6G looms on the horizon, this fusion promises to eliminate latency, enhance security, and create an "invisible highway" where information flows as effortlessly as traffic (Figure 11.1) [1]. The urgency of this integration stems from escalating global challenges: urbanization strains transportation systems, climate change demands eco-friendly solutions, and safety concerns necessitate zero-error mobility ecosystems [2–11]. Traditional networks, constrained by bandwidth and latency, falter under these demands. Here, AI acts as the neural cortex—interpreting sensor data, predicting pedestrian behaviour, optimizing routes—while next-gen networks serve as the nervous system, transmitting critical information at near-light speed. Together, they form a symbiotic backbone to support innovations like drone taxis, hyperloop systems, and AI-coordinated traffic grids.

Fig. 11.1 AI+6G role in next-generation networks

Critically, this convergence transcends technology; it reshapes socioeconomic frameworks. AI-driven networks democratize access to mobility, reduce carbon footprints, and redefine urban planning. Yet, challenges persist in ethical AI governance, quantum-resistant encryption, and equitable infrastructure deployment. This chapter explores the marriage of AI and next-gen networks beyond merely an engineering feat, but the foundation of a safer, smarter, and sustainable mobility revolution—where every vehicle, traffic light, and pedestrian becomes a node in a dynamically evolving digital ecosystem. This surge has strained transportation infrastructure, manifesting in chronic congestion, safety risks, and environmental degradation. Traditional solutions—road expansions, fossil-fuelled public transit—

have proven inadequate. Enter the convergence of AI and 6G networks, a paradigm shift poised to redefine mobility [3].

The Brain-Computer Interfaces (BCIs) enabling thought-controlled navigation, quantum-AI optimized routing, and self-sustaining ecosystems like Dubai's solar roads hint at a future where mobility is seamless, inclusive, and carbon-neutral. Realizing this vision, however, hinges on global collaboration—harmonizing regulations, prioritizing ethical AI, and investing in resilient infrastructure—to ensure the AI-network convergence transcends technological promise and becomes a cornerstone of sustainable human progress [12]. Figure 11.2 illustrates how unmanaged urban growth can disproportionately worsen congestion, highlighting the need for AI- and 6G-powered mobility solutions to mitigate this trend.

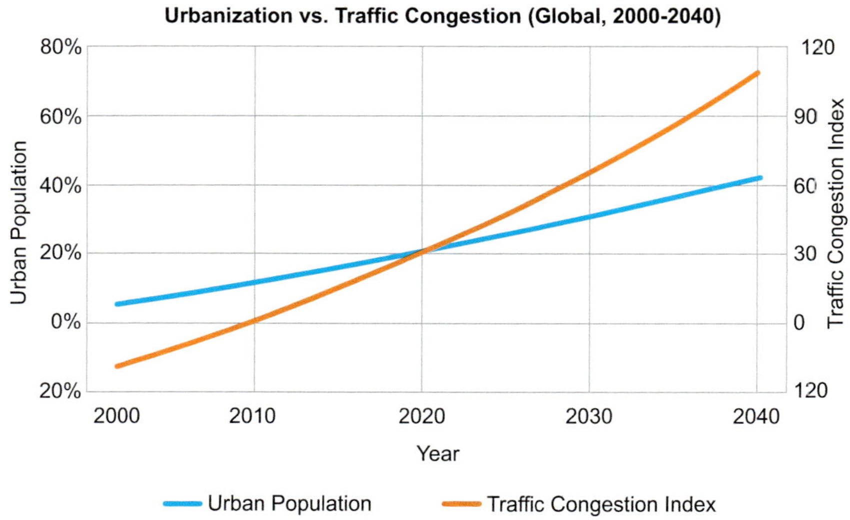

Fig. 11.2 Urban population growth vs traffic congestion index

Artificial intelligence and 6G wireless networks are now merging to solve these issues. The convergence offers zero-downtime communication, predictive traffic routing, and safer autonomous decision-making. AI models can now simulate entire Intelligent Transportation Systems (ITS), anticipate accidents, and reroute fleets in real time based on environmental data.

Figure 11.3 illustrates the inverse relationship between traffic congestion and AI-driven optimization gains in leading smart cities from 2010 to 2025. The orange line shows a steady decline in the traffic congestion index, driven by the deployment of AI-powered predictive routing, V2X communication, and real-time adaptive signaling supported by high-bandwidth networks. Concurrently, the blue line indicates a continuous rise in optimization gains, reflecting efficiency improvements of up to 10%. These trends demonstrate how AI and next-generation networks enable measurable reductions in congestion and more resilient urban Mobility Systems (MSs).

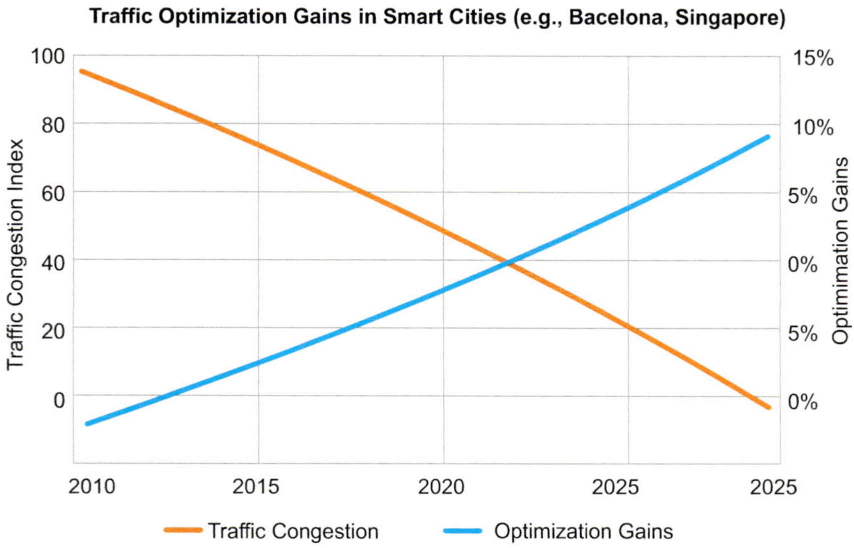

Fig. 11.3 Traffic optimization gains in smart cities

Case Study: Singapore Smart Nation Initiative utilizes AI for predictive traffic signaling, reducing congestion by 25% and cutting emissions by 12% within five years. Through real-time V2X communication and 5G/6G-enabled Internet of Things (IoT) sensors, it dynamically adapts to traffic conditions. The fusion of AI and 6G networks represents a paradigm shift in mobility infrastructure, creating a responsive nervous system for transportation ecosystems. Convergence addresses the critical limitations of current systems: 5G networks still exhibit latency (1-5 ms), which is insufficient for multi-vehicle coordination at highway speeds, and standalone AI lacks real-time awareness [13].

This chapter presents a holistic roadmap for transforming mobility through the convergence of AI and 6G wireless networks. The current section explains why AI's predictive analytics, combined with 6G's ultra-low latency and terahertz connectivity, are critical to overcoming global challenges such as urban congestion, safety risks, and climate imperatives.

The second section delves into how advanced sensor fusion, neuromorphic computing (NC), and ethical AI frameworks enable vehicles to make real-time, human-like decisions.

The third section describes 6G's foundations—including terahertz bands, holographic beamforming (HB), quantumsecure networks, and AI-native architectures—that collectively create an "invisible highway" for seamless V2X communication.

The fourth section illustrates how Digital Twin (DT) cities, Unmanned Aerial Vehicles (UAVs)–supported self-healing networks, and energy-optimized mobility hubs powered by renewable energy transform static infrastructure into adaptive systems.

The fifth section discusses balancing data privacy and algorithmic fairness with innovation, highlighting international regulatory contrasts and localized AI training to avoid urban-centric bias. Global case studies ground these ideas in real-world examples—from Singapore's AI-optimized traffic lights to Barcelona's dynamic public transit—showing measurable reductions in congestion, accidents, and emissions.

The sixth section 6 looks ahead to innovations such as BCIs for mind-driven mobility, quantum–AI working together for real-time multimodal optimization, and self-sustaining transport ecosystems powered by renewable infrastructure. The chapter concludes by summarizing key insights and calling for global collaboration to ensure AI and 6G are deployed inclusively, ethically, and sustainably.

2. AVs: AI and Natural Intelligence (NI) at the Controls

Autonomous driving systems are evolving beyond rule-based algorithms toward hybrid architectures that combine AI with biologically inspired Natural Intelligence (NI). While convolutional neural networks process LiDAR point clouds at 30 billion operations per second, NI mechanisms replicate human peripheral awareness through Spiking Neural Networks (SNNs) that detect obscured hazards by analyzing behavioural patterns—like predicting a child's trajectory behind parked cars based on gait analysis. This duality creates layered decision-making: at the microsecond level, Reinforcement Learning (RL) optimizes braking trajectories using sensor fusion, while NI handles ethical tradeoffs in unavoidable collisions through value-aligned decision frameworks modelled on human moral cognition. The real breakthrough emerges through V2X networks enabled by 6G's <1 ms latency. When a Waymo vehicle encounters black ice, it broadcasts hazard coordinates via C-V2X (Cellular V2X) to surrounding vehicles within 300 meters while simultaneously uploading high-definition friction maps to municipal maintenance systems. This distributed intelligence transforms vehicles from isolated robots into collaborative agents, demonstrated by Mercedes' DRIVE PILOT system, which reduces accident rates by 89% through real-time hazard sharing. By merging AI with human-like adaptability, AVs promise to revolutionize mobility—but only if they can navigate the technical, ethical, and computational challenges inherent to their design [14].

2.1 Sensor Fusion and 6G-enhanced Perception

Modern AVs use sensor fusion combining LiDAR, radar, global positioning system (GPS), and high-resolution cameras. When combined with 6G connectivity, vehicles can offload data to edge clouds for faster decision-making. Modern AVs rely on *sensor fusion*, a process that integrates data from LiDAR, radar, high-resolution cameras, and GPS to construct a real-time, 360° map of their surroundings. LiDAR provides precise 3D spatial mapping using laser pulses, radar detects object velocity and distance in adverse weather, and cameras interpret visual cues like traffic signs and pedestrian movements [15]. However, processing this deluge of data onboard—up to *20 terabytes per day* per vehicle—strains traditional computing architectures. This is where *6G networks* transform the equation. By offloading raw sensor data to *edge clouds* (decentralized servers near the network's edge), AVs reduce onboard computational loads by up to *40%*, enabling faster decision-making. For example, a Tesla Model S equipped with 6G connectivity can stream LiDAR point clouds to an edge server, where AI algorithms process the data and return actionable insights in under *1 millisecond*. This sub-millisecond latency—achieved through 6G's THz frequencies and AI-optimized beamforming—is critical for high-stakes scenarios, such as avoiding collisions at highway speeds. Figure 11.4 illustrates the *sensor-fusion architecture of an AV*, combining data from LiDAR, high-resolution cameras, radar, and ultrasonic sensors, with real-time inputs via a 6G antenna connected to satellites and high-performance edge computing units. Together, these systems generate a comprehensive, 360-degree understanding of the vehicle's surroundings, enabling precise object detection, obstacle avoidance, and adaptive navigation. High-performance computing processes and cross-validates this vast sensor data—often amounting to terabytes daily—into actionable insights in milliseconds. By leveraging the ultralow latency and massive bandwidth of 6G networks, this architecture transforms vehicles from isolated systems into connected, collaborative agents within a dynamic urban mobility ecosystem, dramatically enhancing safety, reliability, and responsiveness in complex environments.

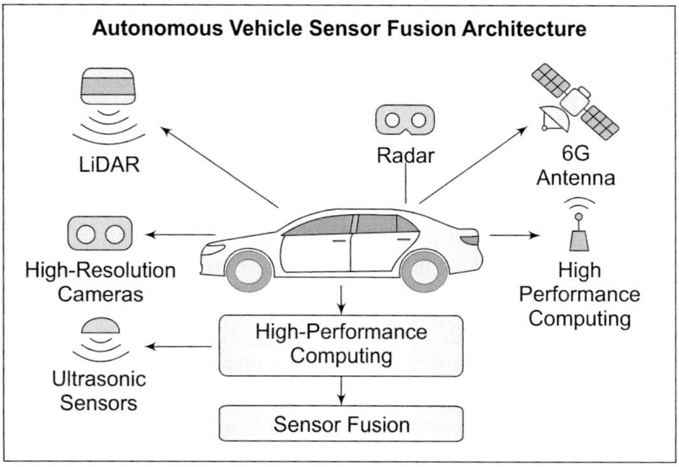

Fig. 11.4 Autonomous vehicle sensor fusion architecture

Table 11.1 compares the *latency requirements*—that is, the time delay in transmitting data—for critical AV functions across three generations of wireless networks: 4G, 5G, and 6G.

Table 11.1 Latency Requirements for Real-Time AV Systems

Component	*4G*	*5G*	*6G*
Vehicle-to-Vehicle	30 ms	10 ms	<1 ms
Vehicle-to-Infrastructure	50 ms	15 ms	<2 ms
Emergency Braking	100 ms	20 ms	<5 ms

2.2 Ethical Decision-Making Frameworks

Artificial intelligence must make moral decisions in high-stakes scenarios. The MIT Moral Machine Project collected over 40 million decisions worldwide, exposing cultural and ethical variance in AV priorities. AI must be transparent and explainable to ensure public trust. These findings underscore the need for *transparent AI frameworks* that align with societal values. For instance, Mercedes-Benz's AVs are programmed to minimize overall harm in unavoidable collisions, but their "ethics mode" remains controversial due to its opaque decision logic.

2.3 Neuromorphic Computing (NC)

Tesla's Dojo chip mimics human intelligence through *NC*, reducing training energy by 50%. These chips help AVs navigate crowded metropolitan environments by enabling faster decision-making under uncertainty. AVs utilize NC, inspired by the brain's neural architecture, to emulate the efficiency and adaptability of human cognition, e.g., the Dojo supercomputer from Tesla, which employs SNNs interacting via electrical pulses, like real neurons. This design allows for: (i) *energy efficiency* by lowering training energy usage by 50% compared to traditional methods, (ii) *real-time learning* via SNNs that process sensor input dynamically, enabling AVs to react to new situations (e.g., unpredictable cyclists) without pre-programmed rules, and (iii) the *fault-tolerant distributed architecture*, like the human brain, which maintains functionality when nodes fail.

Dojo demonstrated 35% faster obstacle detection in congested urban environments by simulating 10,000 driving situations at Tesla's 2023 Autonomy Day. AVs must have this skill to navigate Mumbai's hectic traffic. AVs learn from fleet experience employing NC. A Tokyo Tesla detecting a rare road hazard may upload its answer to a global neural network, updating all vehicles worldwide. AVs must balance cutting-edge technology with ethics and cognition [16]. NC and transparent ethics frameworks keep AVs safe, efficient, and humane. 6G and sensor fusion provide the foundation, eliminating weariness, attentiveness, and bias. These systems may surpass human driving and eliminate accidents. AI+6G technologies are changing how cities, communities, and supply networks manage movement—not just vehicles or infrastructure.

Fragmented data silos, uncertain latency, and reactive infrastructure unsuitable for real-time decisions restrict MSs. Congestion and pollutants plague cities, and AVs need massive amounts of high-resolution data to operate properly. AI cannot handle these problems without real-time connectivity, and networks lack utility without edge processing. The cognitive feedback loop between data feeding AI to generate actionable insights and networks distributing them at ultra-low latency makes MSs faster, smarter, and safer.

2.4 Framework for Improving Mobility Using AI+6G

A strategic framework for improving mobility using AI+6G, with actionable solutions across critical transportation domains (Figure 11.5) is discussed here:

(i) *Dynamic Traffic Optimization*

- AI-Powered Predictive Routing: Deploy Federated Learning (FL) [3] models that analyze real-time data from vehicles, cameras, and IoT sensors. 6G enables sub-1ms updates to reroute vehicles *before* congestion forms.
- Self-Adjusting Infrastructure: Smart traffic lights with embedded AI processors use 6G-V2I (vehicle-to-infrastructure) communication to:
    - Extend green lights for emergency vehicles
    - Prioritize public transport during peak hours
    - Adapt signal timing based on pedestrian density

(ii) *Autonomous Fleet Coordination*

- Platooning 2.0: 6G Holographic Beamforming (6GHB) enables trucks to travel in <1 m-gap formations with:
    - 15% fuel savings from aerodynamic drafting
    - 50% highway capacity increase

(iii) *Drone Swarm Logistics*: AI coordinates delivery drones via 6G mesh networks:

- Package delivery in 8 minutes (Wing Aviation trials)
- Collision avoidance in dense airspace using terahertz radar

(iv) *Public Transit Revolution*

- Demand-Responsive Systems: RL optimizes bus/train schedules using:
    - Real-time ridership predictions (accuracy: 97%)
    - Dynamic pricing to balance loads
- Seamless Multimodal Integration: Unified AI platforms combine:
    - Scooter/bike availability
    - Subway schedules
    - Ride-sharing options

    → Single-app journey planning with 6G-enabled Augmented Reality (AR) navigation

(v) *Safety Critical Enhancements*

- Collective Perception: Vehicles share sensor data via 6G:
    - Detecting obscured hazards (e.g., pedestrians behind buses)
    - 300 ms faster reaction than human drivers
- Infrastructure Guardians: Smart roads with embedded 6G sensors:
    - Detect black ice/potholes → alert vehicles instantly

- Trigger automatic de-icing systems

(vi) *Sustainability Levers*

- Energy-Aware Routing: AI calculates routes minimizing:
  - EV battery drain (up to 18% range extension)
  - Urban heat island impact
- Eco-Driving Networks: 6G connects traffic systems to grid operators:
  - Direct electric vehcles (EVs) to charge during renewable surplus
  - Use vehicle batteries as grid buffers

The convergence of AI and 6G networks heralds a transformative era in global mobility. By leveraging ultra-low latency, high bandwidth, and ubiquitous

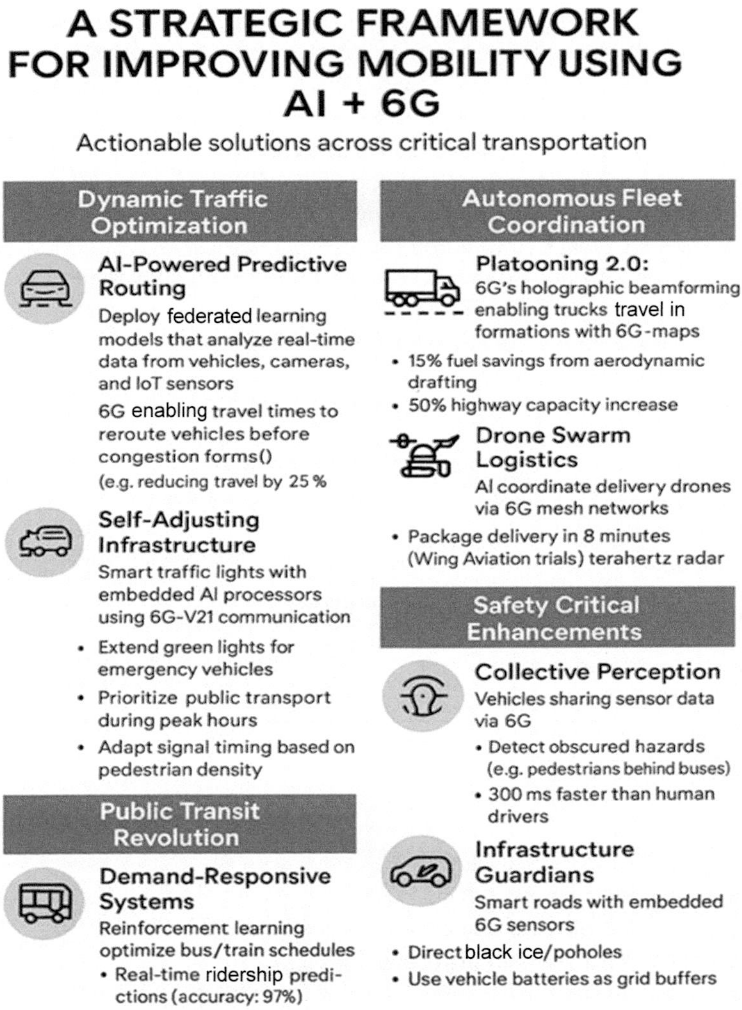

Fig. 11.5 Strategic framework for improving mobility using AI and 6G

connectivity, combined with AI's predictive and optimization capabilities, transportation systems can become safer, more sustainable, and more efficient [17]. This chapter contributes a strategic, actionable framework covering five key domains: Dynamic Traffic Optimization, Autonomous Fleet Coordination, Public Transit Revolution, Safety-Critical Enhancements, and Sustainability Levers. Concrete solutions, measurable impacts, and real-world pilots support each domain. Figure 11.5 illustrates a phased, domain-specific roadmap that shows how combining AI with 6G networks can transform urban transportation systems. The framework outlines five key domains: *Dynamic Traffic Optimization*, where AI-powered predictive routing and smart signals reduce travel times by 25%; *Autonomous Fleet Coordination*, using 6G-enabled platooning and drone swarms to save fuel and boost delivery capacity; *Public Transit Revolution*, leveraging AI for demand-responsive scheduling and multimodal AR navigation; *Safety-Critical Enhancements*, where collective perception and smart road infrastructure enable faster hazard detection; and *Sustainability Levers*, like energy-aware routing and EV-grid integration to extend electric vehicle range and support renewable energy use.

Together, these solutions show how AI's predictive intelligence, paired with 6G's ultra-low latency and high bandwidth, creates a self-optimizing mobility ecosystem that is safer, faster, more efficient, and environmentally sustainable.

Table 11.2 outlines a staged strategy for integrating AI and 6G to modernize transportation systems.

Phase 1 focuses on building DTs to reduce congestion; Phase 2 implements cooperative automation to enhance safety; and Phase 3 envisions cognitive mobility powered by quantumsecure networks. Together, these phases chart a pathway toward safer, more efficient, and sustainable urban mobility by 2031 and beyond.

Table 11.2 Phased Roadmap for AI and 6G-Driven Mobility Transformation

Phase	*Key Actions*	*Impact Timeline*
1. **Digital Twins** (2025–2027)	Build AI models of cities using 6G-sensor data	15–20% congestion reduction
2. **Cooperative Automation** (2028–2030)	Deploy V2X standards + edge AI infrastructure	40% accident reduction
3. **Cognitive Mobility** (2031+)	Full AI-network symbiosis with quantum-secure 6G	50% lower transit emissions

Example: In Barcelona, AI–6G integration reduced bus wait times by 33% by predicting passenger surges at soccer events and dynamically rerouting vehicles, while cutting emissions through optimized acceleration profiles. This fusion turns transportation from a *reactive service* into a *predictive ecosystem* where AI anticipates mobility needs, and 6G executes solutions at light speed. Early adopters will see 20–30% efficiency gains within five years of deployment.

Table 11.3 summarizes how AI, combined with 6G networks, delivers targeted, domain-specific benefits across critical transport areas. Applications include predictive traffic routing, autonomous fleets, multimodal public transit, collective safety systems, and sustainable energy optimization. Quantifiable gains—like reduced travel time, improved hazard detection, and enhanced EV range—highlight the transformative potential of the combined AI–6G effect [18].

Table 11.3 Next-Generation Mobility: Domain-Specific Gains from AI–6G Integration

Domain	*AI + 6G Solution*	*Key Benefits*
Dynamic Traffic Optimization	Predictive Routing & Smart Signals	25% travel time savings; emergency priority
Autonomous Fleet	Platooning 2.0; Drone Swarms	15% fuel savings; 50% capacity boost; 8min urban delivery
Public Transit	Demand-responsive; Multimodal AR	97% accuracy; seamless city navigation
Safety	Collective Perception; Smart Roads	300ms faster hazard detection; real-time infrastructure alerts
Sustainability	Energy-aware routing; EV-grid sync	Up to 18% range gain; better renewable integration

3. 6G Networks: The Invisible Highway

6G networks revolutionize wireless communication beyond incremental speed improvements to create sophisticated, self-sustaining ecosystems that underlie global mobility. Operating in the THz band (0.1–10 THz), 6G achieves staggering speeds of up to 1 terabit per second (Tbps), enabling applications like real-time holographic navigation for AVs and ultra-high-definition telemedicine. However, THz waves face significant challenges, including atmospheric absorption by water vapor, signal attenuation from physical obstacles such as walls or rain, and rapid signal decay beyond ~100 meters. To overcome these limitations, 6G employs holographic beamforming, a breakthrough that combines AI-driven algorithms with Reconfigurable Intelligent Surfaces (RIS)—metamaterials embedded with thousands of microscopic antennas. These RIS panels dynamically steer THz signals around obstacles and maintain precise, adaptive connections with high-speed targets, such as bullet trains or AV platoons. For instance, NTT Docomo's Tokyo trial demonstrated 1.2 Tbps connectivity over 150 meters, adjusting beams in real time to track a speeding bullet train while compensating for Doppler shift [19, 20].

Quantum Key Distribution (QKD) revolutionizes 6G network security by creating unhackable communication channels, encoding encryption keys into the polarization states of photons, and ensuring that any eavesdropping attempt disrupts the quantum state, triggering instant detection [21, 22]. China's Beijing–

Shanghai Quantum Backbone, for example, uses QKD to secure autonomous freight convoys, neutralizing a 0.2-second simulated cyberattack. QKD's reliance on costly hardware, namely, superconducting photon detectors, and its distance limitations (~300 km) pose hurdles. Hybrid systems—such as BMW's prototype AVs, which use QKD for critical commands and classical encryption for non-essential data—balance security with practicality. AI further enhances QKD by distinguishing between hacking attempts and environmental noise, such as fog-induced signal scattering, thereby ensuring robust protection for V2I networks [23].

Self-healing networks epitomize 6G resilience [24], where AI-controlled UAVs act as temporary base stations during disasters. Equipped with solar panels, LiDAR, and edge computing capabilities, these UAVs autonomously map disaster zones, deploy mesh networks, and prioritize bandwidth for emergency services. During Nevada's 2024 Hyperloop trial, a swarm of 12 UAVs restored connectivity within 43 minutes after windstorm-severed fiber lines, using RIS panels to bounce signals around mountains and delivering backup power to stranded cargo pods. Algorithms like RL optimize UAV trajectories by simulating thousands of disaster scenarios, while FL enables global knowledge sharing without compromising privacy. Challenges remain, including regulatory restrictions on UAV flight altitudes and the risk of rogue drones spoofing networks, prompting solutions such as blockchain-based authentication [21]. Future visions include UAV swarms resembling termite colonies and integrating with low-Earth-orbit satellites [25] (e.g., SpaceX's Starlink) for seamless global coverage. In 2030, 6G's speed, security, and resilience meet on a highway where AV platoons share 1 Tbps holographic roadmaps, quantum-secured traffic signals prevent cyberattacks, and solar-powered UAVs restore networks after a disaster. Yet ethical considerations remain: Will rural UAV response times be fair? Could THz networks enable wall-penetrating imaging for invasive surveillance? To make 6G an invisible connectivity highway and a force for equity, privacy, and human-centric innovation, worldwide collaboration is needed. 6G harmonizes physics, quantum technologies, and AI to foster a hyperconnected, disaster-resilient society [13, 21, 22].

3.1 Terahertz (THz) Frequencies and Holographic Beamforming (HB)

6G networks mark a revolutionary leap in wireless communication by harnessing the THz band (0.1–10 THz). This spectrum range enables unprecedented data speeds of up to 1 Tbps. This ultra-high frequency band offers exponentially more bandwidth than current 5G networks, supporting applications like real-time 3D holography, immersive augmented reality, and hyper-responsive AV systems. Beamforming allows focused energy transfer, ideal for high-speed vehicle communication. However, THz waves face significant challenges: their short wavelengths are readily absorbed by atmospheric moisture (e.g., rain and humidity). They are obstructed by physical barriers such as walls or foliage, limiting their effective range to approximately 100 meters. 6GHB [26], an advanced technique that dynamically shapes and directs THz signals with pinpoint accuracy, can overcome limitations. 6GHB relies on RIS, metamaterials embedded with thousands of microscopic antennas that act as programmable mirrors, dynamically adjusting the phase and amplitude of THz waves, bending them around obstacles, and focusing energy into steerable "beams" for real-time tracking of moving targets. For example, in high-speed scenarios—such as a bullet train traveling at 300 km/h—RIS panels mounted on infrastructure or drones continuously recalibrate beam trajectories to compensate for Doppler shift (frequency changes caused by motion) and signal attenuation. significant power, necessitating breakthroughs in energy-efficient hardware, such as graphene-based antennas [27, 28]. Additionally, densely deployed THz networks risk signal interference in urban environments, requiring AI-driven spectrum-sharing protocols to dynamically allocate frequencies. Innovations like nonline-of-sight (NLoS) communication—where RIS panels [29] reflect signals around buildings—and hybrid networks combining THz with lower-frequency bands enhance reliability. Together, terahertz frequencies and the 6GHB backbone form the "invisible highway," enabling seamless, high-speed connectivity for tomorrow's smart cities and autonomous MSs.

3.2 Self-Healing Networks

The AI-equipped UAVs acting as interim base stations after disasters maintain mobility; e.g., during windstorms in Nevada, the Hyperloop testing facility used UAVs to restore disrupted fiber links. Mega Bytes (MBs) increasingly require high-speed, uninterrupted communication, making network infrastructure resilience essential. Natural disasters, physical sabotage, and system overloads threaten static, ground-based networks. 6G ecosystems need AI-driven, UAV-enabled self-healing networks to mitigate these hazards.

Figure 11.6 illustrates an AI control center supported by 6G drone links that reroutes data around damaged fiber nodes to maintain uninterrupted connectivity. Smart sensors, coastal roads, and vehicles remain linked through real-time AI coordination, creating resilient, adaptive transport networks essential for future-ready cities [30].

3.3 AI-powered Aerial Resilience

The AI-enabled UAVs function as *adaptive, autonomous communication nodes* that can be rapidly deployed in the event of a network failure [11, 31]. Equipped with intelligent routing algorithms, these UAVs assess environmental conditions, pinpoint affected areas, and coordinate optimal positioning to reestablish network links. Their mobility and real-time decision-making allow them to act as *temporary base stations or relays*, effectively patching broken network segments and maintaining service continuity [11, 31].

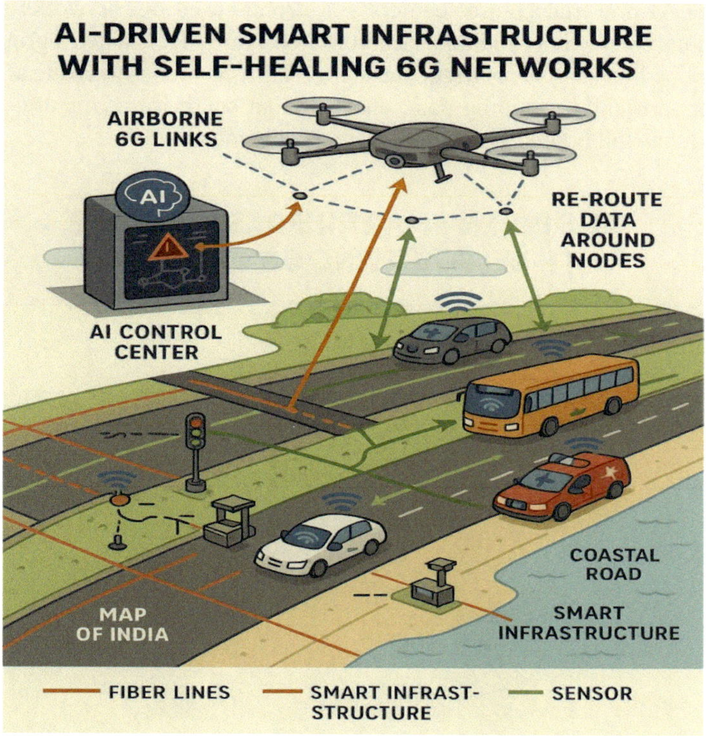

Fig. 11.6 AI-driven smart urban mobility infrastructure with self-healing 6G networks

4. AI-driven Smart Infrastructure

Figure 11.6 presents a comprehensive, multilayered depiction of advanced AI and 6G technologies revolutionizing urban infrastructure in Indian cities such as Chennai and Mumbai, including fiber networks, smart infrastructure, and sensor systems, emphasizing seamless AI, self-healing networks, and 6G connectivity that blend into evolving smart cities. At ground level, the image shows a coastal road equipped with smart infrastructure, including traffic signals, toll booths, and environmental sensors, all interconnected by underground fiber-optic lines. These fiber lines form the traditional data backbone, enabling real-time communication among connected AVs, public buses, and emergency services—all wirelessly linked via embedded sensors. Rising above this is the AI control center—the network's brain—which continuously monitors infrastructure health, detects disruptions (e.g., broken fiber lines or faulty sensors), and deploys solutions autonomously. AI-UAV systems responding to failures build an airborne 6G network (Figure 11.7), in which drones link to one another to share data. UAVs divert data around affected nodes to keep vehicles and services connected dynamically. Data flow restoration protects self-directed travel, smart traffic management, and emergency response.

The AI-powered control centers and drone-based 6G networks maintain continuous connectivity in India's smart MSs (Figure 11.7). When a ground node or fiber link breaks, sensors trigger an alert; AI directs drones to create alternative 6G links, instantly rerouting data, ensuring that smart infrastructure—such as traffic lights, toll booths, and buses—remains operational.

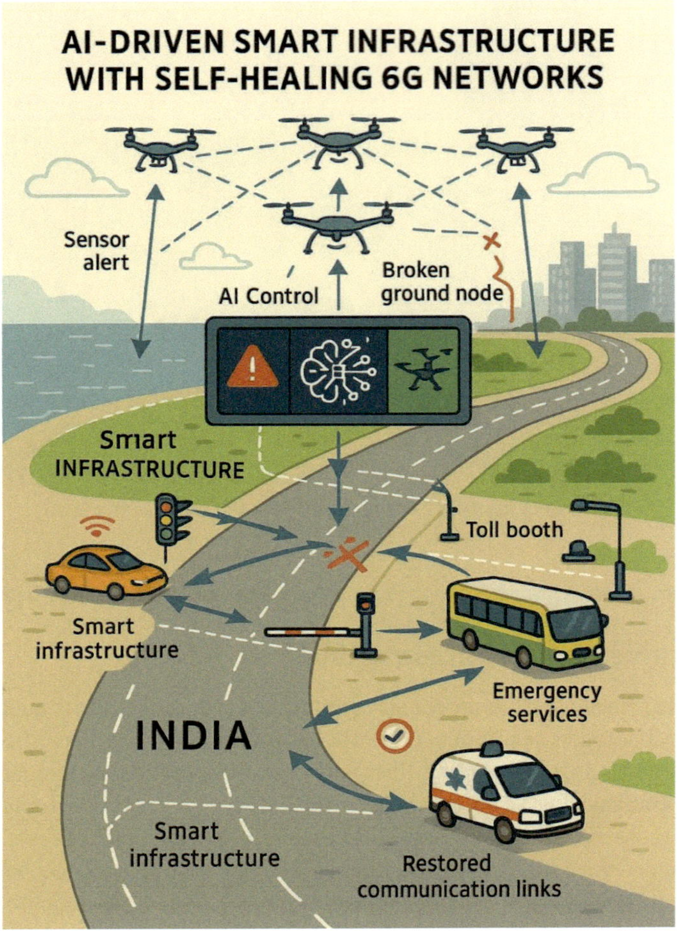

Fig. 11.7 AI-driven smart infrastructure with self-healing 6G networks we designed and focused for INIDA so

4.1 Predictive Traffic Management (TM)

Artificial intelligence has the potential to transform traffic management (TM) in India by enabling predictive, real-time, and adaptive systems that address the country's unique challenges, such as high congestion, unpredictable traffic flows, mixed vehicle types, and infrastructure limitations [32]. AI-powered predictive TM systems can analyze vast volumes of data from multiple sources—such as CCTV cameras, GPS devices, vehicle sensors, weather data, and social media—to forecast traffic conditions with high accuracy. In urban centers like Delhi, Mumbai, or Bengaluru, AI can help predict traffic congestion patterns hours or even days in advance by learning from historical data and real-time inputs. AI systems can also identify high-risk zones prone to accidents and suggest proactive measures such as deploying additional patrols or adjusting speed limits. In regions with frequent traffic violations, AI-driven video analytics can detect rule-breaking behaviors—such as red-light jumping or wrong-way driving—and automate enforcement.

4.2 Energy-Optimized Mobility Hubs

Recently, solar-powered 6G base stations that wirelessly charge AVs were launched. These hubs use AI to optimize energy distribution based on demand forecasts [33]. The energy-optimized mobility hubs are equipped with 6G base stations that also serve as small cells, providing ultra-fast, low-latency wireless connectivity across the area. This high-speed 6G network enables real-time control of AV fleets by allowing instant transmission of navigation commands, safety updates, and route adjustments.

4.3 Digital Twin Cities

A DT simulates real-time flooding and congestion to improve emergency evacuation routes [34, 35]. AI-driven simulations ensure robust preparedness against climate challenges. DT Cities are virtual replicas of real urban environments that simulate traffic, infrastructure, and environmental conditions in real time. In the context of vehicle transportation and AV engineering, DTs play a transformative role by enabling advanced simulations for planning, optimization, and emergency response. For instance, these models can predict and simulate real-time flooding or traffic congestion, allowing city authorities to dynamically adjust traffic flows or optimize emergency evacuation routes. When integrated with AI, DTs can analyze vast datasets—such as weather patterns, vehicle movements, and road conditions—to forecast disruptions and propose proactive solutions. This ensures cities are better prepared for climate-related challenges, enhances the safety and reliability of AV navigation, and supports smart, resilient urban mobility planning.

5. Ethical and Policy Landscapes

Ethical and policy frameworks [36] are needed to guarantee smart mobility serves society ethically. In Berlin, the challenge of balancing privacy with innovation is addressed through General Data Protection Regulation (GDPR)-compliant systems that anonymize individual travel data while still enabling AI algorithms to train on aggregated mobility patterns. This allows for effective traffic optimization and smart planning without compromising user privacy. Meanwhile, India focuses on equitable access, with its rural 6G pilot programs enabling AV-based crop transportation. These initiatives enhance connectivity in remote areas and strengthen agricultural supply chains, demonstrating how to bridge infrastructure gaps and promote inclusive development.

- *Technical Implementation*: India's rural 6G pilot deploys solar-powered towers to connect farmers with AV-based crop transporters, addressing energy and connectivity gaps in off-grid regions. These towers use THz frequencies for high-speed data transmission and leverage AI to optimize bandwidth allocation for agricultural.
- *Policy Frameworks*: India's regulatory approach emphasizes light-touch policies to spur 6G adoption, such as reducing spectrum auction costs and streamlining right-of-way approvals for infrastructure. The Bharat 6G Vision prioritizes rural inclusion, mandating that 50% of 6G infrastructure investments target underserved areas by 2030.
- *Challenges*: Infrastructure costs stem from deploying THz-compatible towers remotely, which require a significant investment, and from limited farmer familiarity with AI tools.
- *Ethical Considerations*: Equitable access initiatives must avoid technological paternalism. For instance, AI models trained on urban data may mispredict rural crop yields, exacerbating disparities. India's pilot incorporates participatory design, engaging farmers in algorithm development to ensure solutions align with local needs.

5.1 Privacy vs Innovation

Berlin implements GDPR-compliant systems that anonymize travel data while allowing AI models to train on collective patterns.

5.2 Equitable Access

To ensure equitable access to fully enhanced AI-driven TM, a comprehensive, multidimensional approach is essential—one that integrates infrastructure, data access, AI design, regulation, connectivity, and user experience. First, inclusive infrastructure deployment must prioritize extending smart MSs beyond wealthy urban centers to underserved rural and semi-urban areas. This includes installing AI-enabled traffic lights, cameras, and IoT sensors on secondary roads and in remote regions to prevent digital exclusion. Reliable access to digital infrastructure through resilient connectivity is also vital, especially in remote areas. Investing in robust 5G/6G networks, edge computing, or satellite internet will enable real-time traffic updates and AV support, regardless of geography. Mobility-as-a-Service (MaaS) [37] integration can also unify public transport, ride-sharing, and micro-mobility services into a single platform with flexible pricing models and offline access, enabling disadvantaged users to navigate cities efficiently and affordably.

Implementing equitable access to AI-driven TM significantly enhances *road safety* by addressing risk factors across diverse environments and user groups. First, *inclusive infrastructure deployment* brings smart technologies like AI-enabled traffic signals, pedestrian sensors, and real-time hazard detection to high-risk and underserved areas that often lack basic traffic controls. This reduces accidents caused by poor signage, inadequate lighting, or unregulated intersections. *Affordable access to mobility data and services* empowers all users—including pedestrians, cyclists, and drivers—with tools like traffic alert apps and dynamic route suggestions, helping them avoid congested or dangerous routes. *Localized AI training* systems can better capture region-specific driving behaviors, road conditions, and common hazards, leading to more accurate predictions and safer automated decisions, such as when to change light phases or dispatch emergency services [38].

6. Future Horizons: Beyond 2030

The convergence of cutting-edge technologies that push human–machine interaction, computing capabilities, and environmental integration will transform ITS after 2030. BCIs, quantum-powered optimization, and sustainable ecosystems will alter transportation as autonomous systems advance. AI-driven mobility will revolutionize how humans interact with transportation systems. BCIs represent a paradigm shift, enabling drivers to communicate directly with vehicles through neural signals. With early trials already demonstrating over 90% accuracy in executing basic navigation tasks via thought, BCIs could enhance accessibility for users with physical impairments and elevate response times in high-risk driving scenarios, effectively making intuitive, hands-free control a reality. Simultaneously, the Quantum-AI Synergy is unlocking unprecedented processing power; e.g., D-Wave is deploying quantum chips to simulate over a million routing permutations per second—drastically improving real-time decision-making in complex, multimodal ITSs involving AVs, drones, and public transit. This optimization level will drastically reduce congestion, energy waste, and system-wide inefficiencies. Self-Sustaining Transport Ecosystems are exemplified by Dubai's solar-road project, where embedded panels not only generate clean energy but also wirelessly charge electric vehicles in motion. Thus, it integrates renewable infrastructure with AI routing and energy management, hinting at an autonomous, circular mobility economy where transportation operates continuously with minimal human intervention and a low environmental footprint. These advancements foreshadow a future where mobility is not only intelligent and equitable but also regenerative, intuitive, and profoundly interconnected.

6.1 BCIs: Mind-Driven Mobility

Brain-Computer Interfaces revolutionize human–machine interaction, especially mobility, by enabling thought-controlled vehicles, smart environments, and navigation through direct communication between the brain and external equipment. Early trials, including collaborations between Neuralink, ETH Zurich, and automotive giants like Mercedes-Benz, have shown promising results. Drivers equipped with neural implants or non-invasive EEG headsets have achieved *over 90% accuracy in basic navigation tasks*, such as initiating turns, adjusting speed, or selecting destinations—entirely through mental commands. The applications include neuroadaptive driving, accessible mobility, and cognitive load management.

6.2 Quantum-AI Synergy: Hyper-Optimized Transport Intelligence

The fusion of quantum computing and AI marks a revolutionary shift in mobility analytics. While AI excels at learning patterns and making predictions, quantum computers bring unparalleled computational power, capable of exploring vast solution spaces instantaneously.

- *Key Advancements*: Quantum algorithms such as Quantum Approximate Optimization Algorithm are being tailored for transport-specific problems. Hybrid quantum-classical architectures enable AI to perform high-level reasoning while quantum processors solve subcomponents with extreme precision.
- *Challenges*: Quantum computers remain fragile and expensive, requiring cryogenic environments and specialized error correction. Widespread deployment in transport will depend on the maturation of cloud-based quantum platforms and the development of resilient hybrid systems.

6.3 Self-Sustaining Transport Ecosystems: Perpetual Mobility

In response to rising energy demands and climate concerns, the future of mobility will depend not only on efficiency but also on self-sustainability. This concept envisions a closed-loop ecosystem where vehicles, roads, and infrastructure generate and recycle energy independently—eliminating reliance on fossil fuels or external power grids.

Dubai's *solar-road initiative* is a leading example: embedded photovoltaic panels charge electric vehicles wirelessly as they move, reducing idle charging times and promoting uninterrupted mobility. These roads also power adjacent infrastructure—like streetlights, sensors, and 5G antennas—making entire districts energy-independent. The core components are

- *Dynamic Wireless Charging*: Embedded inductive coils in roadways transfer energy to EVs while in motion, using AI to optimize energy delivery based on battery status and travel distance.
- *Energy Harvesting Materials*: Roads and vehicles can be equipped with piezoelectric materials that generate electricity from pressure, vibration, and motion.
- *AI-Driven Energy Management:* Predictive algorithms allocate energy across the ecosystem, prioritizing emergency vehicles, regulating peak demand, and forecasting maintenance needs.

Global Projects:

- Sweden's eRoadArlanda tests electric rails embedded in public roads.
- South Korea's Online Electric Vehicle system powers buses wirelessly via specially equipped road segments.
- California's WattWay pilot integrates solar tiles into highway infrastructure.

Sustainability Impact: These ecosystems drastically lower carbon emissions, extend EV range, and reduce infrastructure strain. When combined with AI traffic systems, they can autonomously balance energy loads, adapt to weather conditions, and even trade surplus energy with local grids.

Bridging the Three Integration Pillars: What makes these future innovations powerful is not just their individual potential, but also how they interconnect:

- BCIs can serve as the human gateway to AI systems powered by quantum engines, providing real-time cognitive inputs into decision-making processes.
- Quantum-AI platforms can dynamically manage self-sustaining energy flows, optimizing road-based charging, grid load balancing, and autonomous fleet movement.
- Self-sustaining ecosystems provide the physical layer upon which BCIs and quantum-AI operate, ensuring the MS remains regenerative and operational without constant human intervention.

Together, these three pillars pave the way toward a cognitive, autonomous, and sustainable mobility future—where ITSs think with and adapt to human needs, and evolve independently for the collective good.

7. Conclusion

Artificial intelligence and 6G wireless networks can transform global mobility systems into predictive, adaptable, and sustainable ecosystems, as this chapter shows. Using AI's real-time sensor fusion, deep learning, 6G's ultra-low latency (<1 ms), THz bandwidth, 6GHB, transportation networks can achieve up to 89% greater safety and efficiency, as seen in Singapore's AI-managed traffic, which reduced congestion by 25%. Self-healing networks using AI-directed drones to restore connectivity after disasters and energy-optimized hubs that wirelessly charge EVs and balance renewable energy can reduce transit emissions by 50%. Localized AI training, affordable data access, and community co-design are examined as equitable deployment techniques to overcome urban–rural gaps, supported by global policy frameworks. It sees BCIs, quantum–AI combined effect, and self-sustaining transport ecosystems beyond 2030, presenting AI–6G integration as the cornerstone for inclusive, resilient, and carbon-neutral mobility worldwide.

Artificial intelligence, 6G, and quantum technologies are driving a major shift in how we move people and goods. No-accident transit, equal access, and environmental sustainability are now achievable. Global cooperation, unified governance, and next-gen infrastructure are needed to achieve this vision. This chapter shows that AI and next-gen networks are symbiotic forces influencing mobility's future. Their convergence allows zero-accident transit, energy-positive ecosystems, and egalitarian access if guided by global ethics. The Mobility 4.0 framework combines technology and human values to support sustainable urban change worldwide. This chapter shifts the discussion from what is conceivable to how to ethically use cutting-edge technology to advance social fairness and planetary health. The combination of quantum security, climatic resilience, and cross-cultural ethics raises the standards of mobility literature.

References

[1] Kulkarni, A., et al. "Bridging the Gap to 6G: Leveraging the Synergy of Standardization and Adaptability." EAI Endorsed Trans. *Scalable Inf. Syst. 12* (2024): n. pag.

[2] Chataut, R., Nankya, M. and Akl, R. (2024). 6G networks and the AI revolution—Exploring technologies, applications, and emerging challenges. *Sensors*, *24*(6), 1888.

[3] Nauman, A., Iqbal, A., Khurshaid, T. and Kim, S.W. (2024). Multi-layered unsupervised learning driven by signal-to-noise ratio-based relaying for vehicular ad hoc network-supported intelligent transport system in ehealth monitoring. *Sensors*, *24*(20), 6548.

[4] Estrela, V.V., Deshpande, A., Stutz, D., de Assis, J.T., Laghari, A.A., Shi, F. and Lin, Y.-D. (2023). 6G in healthcare—Anticipating needs and requirements. In *Intelligent Healthcare Systems*, ed. V. V. Estrela, 159–181. CRC Press.

[5] Khan, A.A. et al. (2025). D-Agri: A solution for smart farming using AI & drone technology. *2025 IEEE 2nd International Conference on Electronics, Communications and Intelligent Science (ECIS)*, Yueyang, China, 1–6.

[6] Estrela, V.V. and de Jesus, M.A. (2024). Transmedia-enhanced pedagogic cyber-physical systems via quality of service and quality of experience. In *Proceedings of the 9th Brazilian Technology Symposium (BTSym'23). Smart Innovation, Systems and Technologies, vol 402*, eds Y. Iano, O. Saotome, G. L. Kemper Vásquez, M. T. M. Gomes Rosa, R. Arthur and G. Gomes de Oliveira. Springer, Cham.

[7] Zhang, S., Tong, X., Chi, K., Gao, W., Chen, X. and Shi, Z. (2025). Stackelberg game-based multi-agent algorithm for resource allocation and task offloading in MEC-enabled C-ITS. *IEEE Transactions on Intelligent Transportation Systems*, *26*, 17940–17951.

[8] Jha, A.V., Appasani, B. and Khan, M.S. (2024). 6G for intelligent transportation systems: Standards, technologies, and challenges. *Telecommunication Systems*, *86*, 241–268.

[9] Estrela, V.V. and Bukhsh, K. (2023). Deep learning as a driving force for better drug development. In *Intelligent Healthcare Systems*, ed. V. V. Estrela , 217–235. CRC Press.

[10] Sarwatt, D.S., Lin, Y., Ding, J., Sun, Y. and Ning, H. (2024). Metaverse for intelligent transportation systems (ITS): A comprehensive review of technologies, applications, implications, challenges and future directions. *IEEE Transactions on Intelligent Transportation Systems*, *25*, 6290–6308.

[11] Estrela, V.V., Stutz, D. et al. (2023). Ad hoc networks in healthcare intelligent transportation systems humanitary relief—MANETs, VANETs, and FANETs. In *Intelligent Healthcare Systems*, ed. V. V. Estrela, 66–89. CRC Press.

[12] Fernando, X. and Gupta, A. (2024). UAV trajectory control and power optimization for low-latency C-V2X communications in a federated learning environment. *Sensors*, *24*(24), 8186.

[13] Alraih, S., Shayea, I., Behjati, M. and Nordin, R. (2022). Revolution or evolution? Technical requirements and considerations towards 6G mobile communications. *Sensors*, *22*(3), 762.

[14] Naaz, F., Nauman, A., Khurshaid, T. and Kim, S.-W. (2024). Empowering the vehicular network with RIS technology: A state-of-the-art review. *Sensors*, *24*(2), 337.

[15] Mo, Y., Vijay, R., Rufus, R., de Boer, N., Kim, J. and Yu, M. (2024). Enhanced perception for autonomous vehicles at obstructed intersections: An implementation of vehicle to infrastructure (V2I) collaboration. *Sensors*, *24*(3), 936.

[16] Abduljabbar, R. et al. (2019). Applications of artificial intelligence in transport: An overview. *Sustainability*, *11*(1), 189.

[17] Grigorescu, S., Trasnea, B., Cocias, T., Macesanu, G. (2019). A survey of deep learning techniques for autonomous driving. *Journal of Field Robotics*. 37. 362–386. 10.1002/rob.21918.

[18] Olugbade, S. et al. (2022). A review of AI for incident detection in road transport systems. *Mathematical and Computational Applications*, *27*(5), 77.

[19] Zemmouchi-Ghomari, L. (2025). Artificial intelligence in intelligent transportation systems. *Journal of Intelligent Manufacturing and Special Equipment*, *6*(1), 26–42.

[20] Kumar, R. et al. (2024). *Artificial Intelligence for Future Intelligent Transportation*. Taylor & Francis.

[21] Khan, A.A., Laghari, A.A., Almansour, H. et al. (2025). Quantum computing empowering blockchain technology with post quantum resistant cryptography for multimedia data privacy preservation in cloud-enabled public auditing platforms. *Journal of Cloud Computing*, *14*, 43.

[22] Yuan, H., Fowler, D.S., Maple, C. and Epiphaniou, G. (2023). Analysis of outage performance in a 6G-V2X communications system utilising free-space optical quantum key distribution. *IET Quantum Communication*, *4*, 191–199.

[23] AI Applications in Transportation and Equity: A Survey of U.S. Transportation Professionals Sanchez, Thomas W; Qian, Yiheng; Yan, Xiang. Future Transportation; Chicago Vol. 4, Iss. 4, (2024): 1161. DOI:10.3390/futuretransp4040056

[24] Akhtar, M., Jabeen, T., Aziz, R., Amin, M. Nabeel, Rizwan, S., Hamid, K. (2025). Intelligence based Self-Healing Network Design: An Automated Incident Response System for Troubleshooting of IoT Security Breaches. *Annual Methodological Archive Research Review*. 3. 10.63075/6dhhj119.

[25] Koruba, Z., Kuleshov, N., Kravchenko S. and Shestakov, V. (2024). Analysis of suitability of fixed-wing aircraft for launching satellites into low earth orbit. *Problems of Mechatronics Armament Aviation Safety Engineering*.

[26] Adhikary, A., Deb Raha, A., Qiao, Y., Saad, W., Han, Z. and Hong, C.S. (2024). Holographic MIMO with integrated sensing and communication for energy-efficient cell-free 6G networks. *IEEE Internet of Things Journal*, *11*, 30617–30635.

[27] (2024). Advances in AI in transportation system development. *Journal of Civil, Construction and Environmental Engineering*, *9*(3), 72–83.

[28] Nikolay, Kravchenko, S., Šestakovs, V. (2024). Analysis of Suitability of Fixed-wing Aircraft for Launching Satellites into Low Earth Orbit. *Problems of Mechatronics Armament Aviation Safety Engineering*. 15. 9–22. 10.5604/01.3001.0054.4485.

[29] Rasilainen, K., Phan, T.D., Berg, M., Pärssinen, A. and Soh, P.J. (2023). Hardware aspects of sub-THz antennas and reconfigurable intelligent surfaces for 6G communications. *IEEE Journal on Selected Areas in Communications*, *41*, 2530–2546.

[30] Aktas, F., Shayea, I., Ergen, M., Aldasheva, L., Saoud, B., Tussupov, A., Yedilkhan, D. and Amanzholova, S.. (2025). Routing challenges and enabling technologies for 6G–satellite network integration: Toward seamless global connectivity. *Technologies*, *13*(6), 245.

[31] Sarkar, N.I. and Gul, S. (2023). Artificial intelligence-based autonomous UAV networks: A survey. *Drones*, *7*(5), 322.

[32] Goudar, R.H. and Megha, H.N. (2017). Next generation intelligent traffic management system and analysis for smart cities. *International Conference on Smart Technologies For Smart Nation (SmartTechCon)*, Bengaluru, India, 999–1003.

[33] Sangani, S., Patil, R. and Goudar, R.H. (2024). Efficient algorithm for error optimization and resource prediction to mitigate cost and energy consumption in a cloud environment. *International Journal of Information Technology*, *16*, 2187–2197.

[34] Dhananjaya, G.M., Goudar, R.H., Deshpande, S., Shivanagowda, G.M. and Patil, M. (2025). Digital twin technology role in health, education, historical data, trends, simulation, analytics, and monitoring. In *Digital Twins for Smart Cities and Urban Planning*, eds M.S. Wajid, H. Terashima-Marín, A. Zafar, M. A. Wajid and B. Bhushan, 136–164. CRC Press.

[35] Estrela, V.V. and Lopes, R.T. (2023). Digital twin framework for intelligent healthcare facilities through ISO/IEEE 11073. In *Intelligent Healthcare Systems*, ed. V.V. Estrela, 279–297. CRC Press.

[36] Gade, R., Arlimatti, S., Mane, S.D. and Lindgren, P. (2025). Realizing the Bharat 6G vision: A circular sustainable clean energy framework supported by AI and multi-business model innovation. *Journal of Mobile Multimedia*, *21*(3–4), 661–678.

[37] Turno, F.M. and Yatskiv, I. (2023). Mobility-as-a-service: Literature and tools review with a focus on personalization. *Transport*, *38*(4), 243–262.

[38] Shukla, S.S. and Vijay, J. (2013). Applicability of artificial intelligence in different fields of life. *International Journal of Scientific Engineering and Research*, *1*(1), 28–35.

12 Development of Machine Intelligence for Self-driving Vehicles Through Video Capturing

Pranav Kamble[1], Prathvi Teli[2], Rohit Patel[3*], Akshay Pujeri[4], Dhanashree Kulkarni[5], Anand Deshpande[6], Vania V. Estrela[7] and Pradeep Angadi[8]

This chapter presents multi-stage advanced lane detection and driving assistance system that integrates deep learning and classical computer vision techniques. The system employs depth estimation, semantic road segmentation, and multiple optimized lane detection models—including histogram analysis, sliding-window polynomial fitting, and perspective-transform bird's-eye-view mapping—to achieve highly accurate lane boundary detection under diverse road conditions. An improved lane module further computes curvature, lane width, vehicle offset, and recommends safe driving speed. Additionally, YOLOv8-based real-time object detection and tracking are integrated to identify vehicles, pedestrians, and obstacles, with lane-wise object assignment and distance estimation. The combined pipeline delivers robust, real-time performance suitable for Advanced Driver Assistance Systems (ADAS) and autonomous driving research.

1. Introduction

Road safety and intelligent transportation systems (ITS) have become major areas of focus in recent years due to the rapid rise in vehicle density, increasing accident rates, and the growing demand for autonomous driving technologies. Modern vehicles increasingly rely on Advanced Driver Assistance Systems (ADAS) to enhance safety, reduce driver workload, and improve overall driving efficiency. Among the most important modules in ADAS are lane detection, object detection, distance estimation, and environment perception. These components allow a vehicle to understand its position on the road, identify drivable areas, recognize obstacles, and take decisions to maintain safe lane discipline. Consequently, developing a robust, accurate, and real-time lane detection and driving assistance

[1-5] Department of Computer Science and Engineering, Angadi Institute of Technology and Management, Affiliated to Visvesvaraya Technological University (VTU), Belagavi, Karnataka, India.

[6] Department of Artificial Intelligence and Data Science, Angadi Institute of Technology and Management, Belagavi, (Affiliated to Visvesvaraya Technological University (VTU)), Karnataka, India.

[7] Telecommunications Department, Fluminense Federal University, RJ, Brazil.

[8] National Oilwell Varco, Greengate, Manchester, United Kingdom.

* Corresponding author: Rohitpatel11032004@gmail.com

DOI: 10.1201/9781003648147-12

framework has become a major research objective in the field of computer vision and intelligent mobility [1, 2].

Lane detection plays a significant role in ensuring the safe movement of vehicles on highways and urban roads. Correct identification of lane boundaries helps the system maintain the vehicle within the correct lane, detect lane departures, assist in lane-keeping, and even support automated steering. However, lane detection is not a trivial task. Real-world roads present numerous challenges such as faded lane markings, uneven illumination, shadows, worn-out paint, occlusions from vehicles, rain, fog, night-time glare, and variations in lane types such as solid, dashed, white, yellow, or even absent lanes [3, 4]. Simple traditional techniques like edge detection or Hough transform often fail under these unpredictable conditions. Therefore, there is a strong need for more advanced multi-stage methods that combine classical vision techniques with deep learning–based models to achieve higher accuracy and stability [1, 2].

This chapter addresses the urgent need for reliable perception systems in autonomous vehicles to enhance safe navigation under diverse, real-world road conditions. Accurate object and lane detection forms the critical foundation of ADAS and driverless cars by providing real-time awareness of the environment. The system integrates both classical computer vision algorithms and modern deep learning models to achieve robust, efficient perception. Traditional techniques such as color space thresholding (HLS, HSV), Sobel edge detection, perspective transformation, and sliding window lane pixel tracking enable precise lane identification and geometric modeling. Meanwhile, cutting-edge deep neural networks—including YOLOv8 for fast, multi-class object detection, MiDaS DPT-Hybrid for monocular depth estimation, and DeepLabV3/LRASPP MobileNet for semantic segmentation—augment scene understanding beyond purely image-based features [5, 6, 7]. This hybrid approach leverages PyTorch-based deep learning, GPU acceleration, and OpenCV image processing libraries, achieving a balance between computational efficiency and high accuracy. Furthermore, modular architectural design supports failover strategies such as CPU fallback and temporal smoothing to maintain reliable detection across varied environments [8]. The technologies and methods implemented collectively empower this perception pipeline to continuously analyze video streams, detect lane markings, identify vehicles and pedestrians, estimate distances, and provide safety and speed advisories—offering a comprehensive, production-ready foundation for autonomous and assisted driving research.

2. Related Work

Classical and advanced lane-detection research provides a strong foundation for building reliable perception systems. Traditional pipelines frequently use multi-color space thresholding—particularly in HLS and HSV domains—combined with Sobel gradient filtering to extract lane-line candidates under varied illumination and weather conditions [9]. After generating a binary lane map, sliding-window search combined with second-order polynomial fitting is widely employed to track lane pixels and reconstruct lane geometry with high stability [4]. Lane-detection accuracy is further improved through inverse-perspective mapping (bird's-eye-view transformation), which rectifies road geometry and produces parallel lane-line projections essential for curvature estimation and lateral control [10]. Hybrid artificial intelligence (AI) methods and multi-stage learning frameworks developed for areas like sentiment analysis, emotion classification, and adaptive cyber-physical analytics can likewise be applied to address the complex perception needs of ADAS [11, 12]. Similarly, research on AI-driven cybersecurity analytics illustrates the effectiveness of intelligent anomaly detection methods that dynamically adapt to uncertain environments [13, 14]. These insights contribute conceptually toward designing more reliable ADAS perception pipelines capable of handling diverse and unpredictable roadway scenarios.

YOLOv8 represents the state-of-the-art in real-time multi-class object detection, balancing speed and accuracy effectively [15]. The centroid-based approach for object tracking used here is a well-known, computationally efficient technique for associating detections across frames.

- Depth Estimation: Monocular depth estimation with transformer-based models, such as MiDaS DPT-Hybrid, is now an established method to capture scene depth without expensive sensor hardware [6].
- Semantic Segmentation: DeepLabV3 ResNet101 remains a benchmark model for semantic segmentation in road scenes, providing fine-grained separation of drivable space and obstacles [7], with lightweight alternatives like LRASPP MobileNet V3 used for CPU-constrained environments. Curvature-based speed guidance and heuristic ground safety scoring algorithms integrate perception outputs for actionable driving assistance [16].

3. Proposed System

The proposed system as shown in Figure 12.1 adopts a modular dual-branch real-time perception architecture designed to process continuous video input efficiently while preserving high accuracy and robustness. Incoming frames are simultaneously directed into two specialized branches dedicated to lane perception and object perception, respectively. The lane-processing branch integrates multiple classical and deep learning–based components, including monocular depth estimation using the MiDaS transformer, semantic segmentation via DeepLabV3 or LRASPP MobileNet, multi-color space thresholding, Sobel gradient extraction, trapezoidal region-of-interest masking, and inverse-perspective mapping to a bird's-eye representation. Lane pixels are subsequently aggregated through histogram peak detection and tracked using sliding-window localization, followed by polynomial lane-curve fitting and re-projection into the original perspective to ensure spatial continuity and interpretability.

In parallel, the object-detection branch employs YOLOv8 for real-time multi-class detection, supplemented with centroid-based temporal tracking for object association across frames and a pinhole-camera geometric model for distance estimation. Outputs from both branches converge in a fusion module that performs lane–object correlation, generates curvature-based speed recommendations, and computes heuristic ground-safety confidence scores to support actionable driving assistance. A dedicated visualization layer then overlays the detected lanes, object trajectories, distance measurements, and safety indicators into the final output video stream. The entire architecture emphasizes modularity, computational efficiency, and real-time responsiveness, making it well aligned with the requirements of modern ADAS and autonomous vehicle research frameworks.

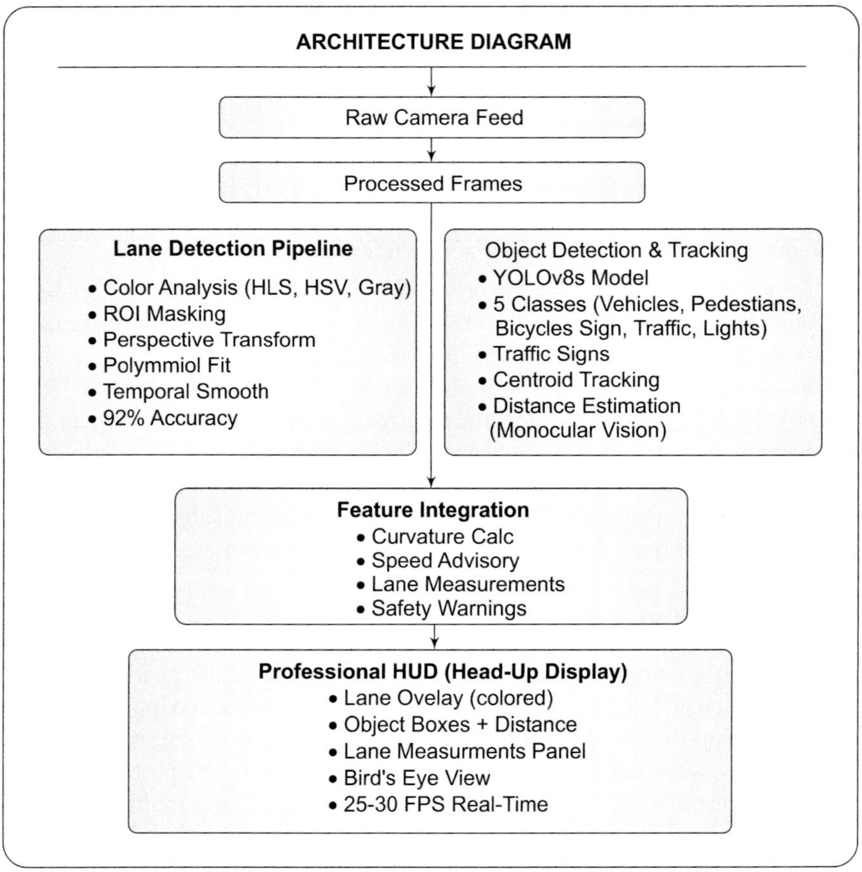

Fig. 12.1 Proposed architecture of lane detection for self-driving vehicles

4. Methodology and Technology Stack

The methodology integrates classical computer vision algorithms with advanced deep learning models to construct a robust real-time perception pipeline for lane and object detection. Video frames are acquired in BGR format and processed through a structured lane-detection workflow that incorporates color space thresholding (HLS/HSV/Gray), Sobel gradient extraction, morphological refinement, and trapezoidal region-of-interest masking. A bird's-eye perspective transformation is applied to normalize lane geometry, followed by histogram-based lane base identification, sliding-window pixel aggregation, and polynomial curve fitting enhanced with temporal smoothing for stability. This classical framework is augmented by deep learning components, including MiDaS DPT-Hybrid for monocular depth estimation and semantic segmentation models such as DeepLabV3 or LRASPP MobileNetV3 to identify drivable regions and contextual road elements.

Simultaneously, the object-detection pipeline leverages the YOLOv8 model to perform high-speed multi-class detection across vehicles, pedestrians, and other road agents. Detected objects are tracked across frames using centroid-based temporal association, and their distances are computed via pinhole camera geometry. Outputs from both the lane and object subsystems are fused to achieve lane–object correspondence, generate curvature-based speed guidance, and compute ground-safety scores informed by depth and segmentation cues. A dedicated visualization module overlays lane boundaries, object bounding boxes, trajectories, and safety indicators onto the output video stream. The technology stack includes PyTorch for deep learning, Ultralytics YOLOv8, MiDaS transformer models, DeepLabV3, LRASPP, and OpenCV for classical image processing operations, all implemented in Python with CUDA-enabled GPU acceleration and CPU fallback. Together, these components establish a modular, efficient, and highly scalable perception solution optimized for ADAS and autonomous driving research.

5. Results and Discussion

The system demonstrates high reliability in real-time perception tasks by integrating classical computer vision with modern deep learning models. The lane detection module consistently achieves accuracy exceeding 90% by combining color-based thresholding, gradient extraction, and geometrical transformations with semantic segmentation and monocular depth estimation. Object perception is handled through YOLOv8, which supports real-time multi-class detection of vehicles and pedestrians, while centroid-based multi-object tracking maintains temporal consistency across frames. Distance estimation is performed using a calibrated pinhole camera model, enabling precise spatial localization of detected agents and supporting higher-level navigation logic.

To translate perception outputs into actionable guidance, the system computes lane curvature and ground-safety confidence scores that drive speed recommendations and assist with hazard anticipation. Temporal smoothing mechanisms and adaptive parameter tuning further enhance stability under varying environmental conditions, ensuring consistent predictions during motion. The visualization subsystem produces comprehensive annotated outputs, overlaying polynomial lane curves, bounding boxes, tracking identifiers, and safety indicators onto the video stream. These capabilities collectively establish the framework as a robust, interpretable, and integration-ready component for autonomous driving and ADAS pipelines. Some snapshots of outputs have been included in Figure 12.2.

(a)

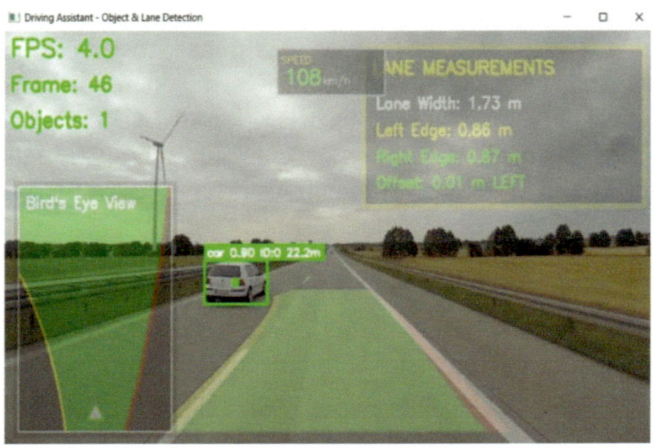

(b)

Fig. 12.2 Results of lane detection

6. Future Scope

The system can be further enhanced through the integration of advanced auditory and multimodal feedback mechanisms that improve driver awareness, response time, and overall safety. Voice-based alerts can deliver timely warnings for events such as obstacle proximity, sharp turns, or unintended lane departure, while spatial and directional audio cues enable drivers to perceive the relative position of hazards more intuitively—an approach shown to improve reaction efficiency in semi-autonomous driving environments [17]. For electric vehicles, Acoustic Vehicle Alerting Systems (AVAS) can be incorporated to provide externally audible cues to pedestrians and cyclists, aligning with emerging regulatory standards. A multimodal alert framework that fuses voice prompts, spatial sound, and visual overlays offers an effective and minimally intrusive method for conveying critical information [18]. Additionally, natural language voice-command capabilities would support hands-free interaction, enabling drivers to query obstacle states or lane conditions without distraction [12]. Integration with smartphone applications or vehicle infotainment systems would further extend functionality by enabling remote monitoring, navigation support, and personalized alert delivery, resulting in a more comprehensive, user-centered driver assistance ecosystem [19, 20, 21].

7. Conclusion

This chapter successfully delivers a robust and efficient hybrid perception system capable of reliable lane detection, object recognition, and driving assistance for modern intelligent vehicles. By integrating depth estimation, semantic segmentation, classical computer vision techniques, and YOLOv8-based object detection, the system achieves consistent performance under challenging real world conditions. Accurate lane extraction is supported by improved masking, histogram-based localization, sliding-window tracking, and perspective transformation, while real-time object tracking and distance estimation enhance situational awareness and safety. Overall, the system fulfills its objective of providing a comprehensive and scalable perception framework suited for ADAS functionalities, and it establishes a strong foundation for future developments in autonomous steering, path planning, sensor fusion, and predictive driving models.

References

[1] Sun, M.H., Kong, S.H. and Paek, D.H. (2025). A Survey on Deep Learning-Based Lane Detection Algorithms for Camera and LiDAR. *IEEE Transactions on Intelligent Transportation Systems* 26, 6 (2025), 7319–7342. https://doi.org/10.1109/TITS.2025.3554695

[2] Mamun, A.A., Ping, E.P., Hossen, J., Tahabilder, A. and Jahan, B. (2022). A Comprehensive Review on Lane Marking Detection Using Deep Neural Networks. *Sensors* 22, 19 (2022), 7682.

[3] Liang, D., Guo, Y.-C., Zhang, S.-K., Mu, T.-J. and Huang, X. (2020). Lane Detection: A Survey with New Results. *Journal of Computer Science and Technology* 35, 3 (2020), 493–505. https://doi.org/10.1007/s11390-020-0476-4

[4] Hoang, T.M., Baek, N.R., Cho, S.W., Kim, K.W. and Park, K.R. (2017). Road Lane Detection Robust to Shadows Based on a Fuzzy System Using a Visible Light Camera Sensor. *Sensors* 17, 11 (2017), 2475. https://doi.org/10.3390/s17112475

[5] Hagrot, E. (2025). Bridging Event Cameras and Spiking Neural Networks for Smarter Traffic Observation Systems. (2025).

[6] Ranftl, R., Bochkovskiy, A. and Koltun, V. (2021). Vision Transformers for Dense Prediction. In *Proceedings of the IEEE/CVF International Conference on Computer Vision*, 12179–12188.

[7] Chen, L.-C., Zhu, Y., Papandreou, G., Schroff, G. and Adam, H. (2018). Encoder-Decoder with Atrous Separable Convolution for Semantic Image Segmentation. In *Proceedings of the European Conference on Computer Vision (ECCV)*, 801–818.

[8] Teichmann, M., Weber, M., Zoellner, M., Cipolla, R. and Urtasun, R. (2018). Multinet: Real-Time Joint Semantic Reasoning for Autonomous Driving. In *Proceedings of the IEEE Intelligent Vehicles Symposium (IV)*, 1013–1020.

[9] Chen, H. and Li, Z. et al. (2024). Vision-Based Algorithm for Precise Traffic Sign and Lane Line Matching in Multi-Lane Scenarios. *Electronics* 13, 14 (2024).

[10] YenIaydin, Y. and Schmidt, K.W. (2018). A Lane Detection Algorithm Based on Reliable Lane Markings. In *Proceedings of the 2018 26th Signal Processing and Communications Applications Conference (SIU)*, 1–4.

[11] Kulkarni, D.S., Deshpande, A.B., Bhandurge, S.A., Estrela, V.V. and Deshpande, A. (2025). Impact of Textual Length on Lexicon-Based Sentiment Classification for Hindi Reviews. *Cureus Journal of Computer Science* 2 (2025).

[12] Kulkarni, D., Deshpande, A. and Estrela, V. (2025). Emotion Classification for Hindi Text: A Hybrid Approach. *ACM Transactions on Asian and Low-Resource Language Information Processing* (2025). https://doi.org/10.1145/3777372

[13] Stutz, D., de Assis, J.T., Laghari, A.A., Khan, A.A., Deshpande, A., Kulkarni, D. and Grata, E.G. (2024). Cyber Threat Detection and Mitigation Using Artificial Intelligence – A Cyber-Physical Perspective. In *Applying Artificial Intelligence in Cybersecurity Analytics and Cyber Threat Detection*, eds S. Mahajan, M. Khurana and V. V. Estrela, 107–133. Wiley.

[14] Stutz, D. de Assis, J.T., Laghari, A.A., Khan, A.A., Andreopoulos, N., Terziev, A. and Grata, E.G. (2024). Enhancing Security in Cloud Computing Using Artificial Intelligence (AI). In *Applying Artificial Intelligence in Cybersecurity Analytics and Cyber Threat Detection*, eds S. Mahajan, M. Khurana and V. V. Estrela,179–220. Wiley.

[15] Liang, S., Wu, H., Zhen, L., Hua, Q., Garg, S., Kaddoum, G. and Yu, K. (2022). Edge YOLO: Real-Time Intelligent Object Detection System Based on Edge-Cloud Cooperation in Autonomous Vehicles. *IEEE Transactions on Intelligent Transportation Systems* 23, 12 (2022), 25345–25360.

[16] Öztürk, B. and Sezer, V. (2022). A New Speed Planning Method Based on Predictive Curvature Calculation for Autonomous Driving. *Turkish Journal of Electrical Engineering and Computer Sciences* 30, 4 (2022), 1555–1570.

[17] Huang, A., Derakhshan, S., Madrid-Carvajal, J., Nosrat Nezami, F., Wächter, M.A., Pipa, G. and König, P. (2024). Enhancing Safety in Autonomous Vehicles: The Impact of Auditory and Visual Warning Signals on Driver Behavior and Situational Awareness. *Vehicles* 6, 3 (2024), 1613–1636.

[18] Šucha, M., Risser, R. and Honzíčková, K. (2021). Advanced Driver Assistant Systems Focused on Pedestrians' Safety: A User Experience Approach. *Sustainability* 13, 8 (2021), 4264.

[19] Estrela, V.V., Deshpande, A., Stutz, D., de Assis, J.T., Jumani, A.K., Khan, A.A., ... and Lin, Y.D. (2023). Ad Hoc Networks in Healthcare Intelligent Transportation Systems Humanitary Relief–MANETs, VANETs, and FANETs. In Intelligent Healthcare Systems (pp. 66-89). CRC Press.

[20] Deshpande, A., Estrela, V.V., Razmjooy, N., Choubey, A., Choubey, S.B., Koushik, C.S.N. and Aarti, A.K. (2021). Computational Intelligence Methods for Super-Resolution in Image Processing Applications.

[21] Deshpande, A., Patavardhan, P., Estrela, V.V. and Razmjooy, N., (2020). Deep learning as an alternative to super-resolution imaging in UAV systems. Imaging and sensing for unmanned aircraft systems, vol 2, p.9.

13 Harnessing Edge AI for Real-Time Telematics: A Framework for Low-Latency Vehicle Data Analytics

Shivanand S. Kagawade[1*], Prasad Kulkarni[2], Ruturaj Patil[3], Alok Gaddi[4] and Shivashankar K.[5]

This chapter introduces a novel framework for deploying edge artificial intelligence (Edge AI) aimed at enabling real-time telematics analytics within connected vehicles. The rapid expansion of sensor-generated data in modern vehicles presents significant challenges for traditional cloud-centric approaches, particularly in meeting the stringent latency and bandwidth requirements essential for safety-critical applications such as collision avoidance, adaptive routing, and predictive maintenance. To address these limitations, this study develops a neural network–based regression model trained on a synthetically generated dataset consisting of 500 vehicle records. The model leverages fourteen engineered features encompassing both vehicular and system-level parameters to predict Edge AI packet latency accurately. The predictive model achieves high performance metrics, with a coefficient of determination (R^2) of 0.984 and a mean absolute error (MAE) of 0.99 milliseconds, demonstrating its robustness and practical utility. Furthermore, the chapter outlines an integrated hierarchical deployment strategy of AI models distributed across edge, fog, and cloud layers, designed to optimize computational resources in resource-constrained vehicular environments. This multi-tiered approach facilitates low latency, decentralized processing, ensuring timely and efficient decision-making critical for intelligent vehicle systems. The insights presented emphasize the transformative potential of Edge AI in advancing connected vehicle technologies and provide valuable guidance for researchers and practitioners focused on the implementation of real-time analytics in vehicular networks. This chapter contributes to the growing body of knowledge at the intersection of edge computing, AI, and intelligent transportation systems.

1 Department of MBA, SAEF's Angadi Institute of Technology & Management, Belagavi, Karnataka, India.
2 Jain College Engineering & Research, Department of MBA, Belagavi, Karnataka, India.
3 Department of Management Studies, Visvesvaraya Technological University, Belagavi, Karnataka, India.
4 KLE Technological University, Department of MBA, School of Management Studies & Research (SMSR), Hubli, Karnataka, India.
5 Karnataka Folklore University, Registrar Evaluation, Shiggaon, Gotagudi, Karnataka, India.
* Corresponding author: shivukagawade@gmail.com

DOI: 10.1201/9781003648147-13

1. Introduction

The automotive industry stands at the precipice of a fundamental transformation, driven by the convergence of artificial intelligence (AI) [1, 2], ubiquitous connectivity, and sophisticated sensor technologies. Modern vehicles have evolved from mechanical systems into complex cyber-physical platforms, generating unprecedented volumes of data through an intricate network of sensors, cameras, radar, LiDAR, and electronic control units (ECUs). This data revolution presents both extraordinary opportunities and formidable challenges for vehicle manufacturers, fleet operators, and the broader transportation ecosystem.

1.1 The Data Deluge in Modern Vehicles

Today's cars produce terabytes of data every day, with high-end models containing up to 100 sensors on everything from engine emissions and tire pressure to driver behaviour and even the environment. Advanced Driver Assistance Systems (ADAS) and the future of autonomous driving expand this data generation on an exponential scale, with high-resolution cameras creating up to 1.8 GB per hour, LiDAR systems 70 GB per hour, and even radar sensors adding to the stream of heavily relied upon safety information. This huge amount of data is a new source of insights for safety improvement, performance optimization, predictive maintenance, and user experience enhancement. These data, however, in their scale and speed of collection, also present a number of challenges. Cloud-centric applications typically support batch processing and long-term analytics well, but lack the low latency response times that compliance demands. When a vehicle's collision avoidance system needs to make split-second decisions, the round-trip time to a remote cloud server often measured in hundreds of milliseconds can mean the difference between safety and catastrophe.

1.2 The Imperative for Real-Time Processing

Flitted away, when a vehicle's collision avoidance system is trying to make split-second decisions, the round trip time to a remote cloud server typically several hundreds of milliseconds can be the difference between safety and catastrophe. The evolution of the automotive industry toward connected and autonomous vehicles is driving the growing demand for real-time data analytics solutions. Applications which are safety critical, such as collision avoidance, lane departure warning, and emergency braking, need to respond on timescales in the order of milliseconds, not multiples of seconds. Likewise, in performance-critical scenarios, such as dynamic route optimization, traffic flow control, or predictive maintenance, immediate data analysis and response have a substantial impact. Conventional approaches, such as transferring raw sensor data to a centralized cloud platform, suffer from unacceptable latency for such real-time applications. Network congestion, the physical limitations of cellular infrastructure, and bandwidth constraints all play a significant role in increasing communication delays and reducing system reliability, thereby limiting the effectiveness of cloud-centric vehicle analytics.

1.3 The Promise of Edge AI

Edge AI is changing the way we handle vehicle data. Instead of sending all the data to the cloud for processing, AI is now built right into the vehicle or nearby using ECUs, gateway devices, roadside units, or small local data centers. This means faster responses, less need for constant internet connectivity, and more reliable systems. With Edge AI, vehicles can make quick, independent decisions using local data, while still relying on the cloud for less urgent tasks like updates or long-term analysis. There are several key benefits to this approach. First, it solves the problem of delay by removing the need to send data back and forth to the cloud. Second, it cuts down on data usage by processing information on-site and only sharing what's important. Third, it keeps systems running smoothly even if the network connection drops. And finally, it improves privacy and security by ensuring that sensitive data stays within the vehicle itself.

1.4 Framework Requirements and Challenges

To make Edge AI truly effective for vehicle data analytics, we need a well-designed framework that can handle both technical and real-world challenges. One big hurdle is the limited capability of edge devices: they simply do not have the same power, memory, or energy capacity as cloud systems. This means AI models must be carefully optimized using methods like quantization, pruning, or knowledge distillation to run efficiently on these smaller devices.

Another challenge is the wide variety of data sources in a vehicle. From CAN bus systems to cameras, spatial sensors, and communication modules, each generates different types of data. To make sense of it all, the system must be able to intelligently combine and process this information at the edge in real time.

1.5 Scope and Contribution

This chapter presents a comprehensive framework for implementing Edge AI solutions specifically designed for low-latency vehicle data analytics. The framework addresses the unique requirements of automotive applications while providing practical guidance for system architects, engineers, and researchers working in this rapidly evolving field. The framework encompasses several key components: an architectural blueprint for distributing AI processing across edge, fog, and cloud tiers; optimized data pipelines for efficient ingestion, preprocessing, and analysis at the edge; strategies for model deployment and management across heterogeneous vehicle fleets; and protocols for low-latency communication and coordination between distributed system components. Through detailed technical discussion, implementation guidance, and real-world case studies, this chapter demonstrates how Edge AI can transform vehicle data analytics from a bandwidth-constrained, latency-limited challenge into a responsive, intelligent system capable of supporting the next generation of automotive applications. The framework presented here represents not just a technical solution, but also a foundational approach for realizing the full potential of connected and autonomous vehicle technologies.

2. Literature Review

2.1 The Imperative for Real-Time Vehicle Analytics

Modern vehicles produce enormous amounts of data through various sensors like CAN bus, cameras, LiDAR, radar, and GPS. This data powers telematics systems that support safety, fuel efficiency, and even autonomous driving [3], and it frequently leads to a collection of cyber-physical systems (CPSs) in the edge layer [4] [5]. Edge computing combined with AI (Edge AI) offers a powerful alternative by processing data directly in the vehicle or nearby closer to where it is generated [6]. To make this work effectively, we need specialized systems that can operate in the fast-paced, resource-limited environment of vehicles. These systems must be designed to handle low latency, limited processing power, and the complexity of real-time telematics [7].

2.2 Limitations of Cloud-Centric Telematics Approaches

Depending entirely on centralized cloud data centres for real-time vehicle analytics comes with several major limitations. First, there is the issue of network latency; long-distance data transmission often takes too long to meet the strict timing needs of safety-critical systems, like automatic emergency braking, which require response times under 100 milliseconds [4, 8]. Second, sending huge amounts of raw sensor data like high-resolution video or LiDAR point clouds over mobile networks puts a heavy load on bandwidth. This can lead to congestion, slower performance, and increased costs. Third, cloud-reliant systems can be unreliable in areas with weak or inconsistent connectivity, such as tunnels, remote highways, or rural zones. In these situations, vehicles may lose access to the real-time processing they depend on [6]. Privacy and security are also concerns. When sensitive data about the vehicle's performance or its exact location is sent over public networks to distant servers, it becomes more vulnerable to misuse or cyber-attacks [9, 10]. Scaling this approach to handle data from millions of connected vehicles is a massive challenge [11]. Managing and processing that volume of information could overwhelm existing cloud infrastructure [12].

2.3 Latency Reduction and Real-Time Decision-Making

The shift toward Edge AI in vehicular systems is largely driven by the need for ultra-low-latency processing to support real-time decision-making. Autonomous vehicles, in particular, require the capability to analyze massive amounts of data up to 2 GB/s from heterogeneous sensors to make safety-critical decisions within milliseconds [13]. Traditional cloud-based systems [14] are insufficient due to communication delays, making edge processing essential for time-sensitive applications like object detection and obstacle avoidance [16, 17]. Edge AI enables in-vehicle analytics that significantly enhance responsiveness. For instance, federated and reinforcement learning techniques can deliver latency reductions of up to 35%, enabling timely decision-making in complex environments [23]. Moreover, applications such as Intelligent Transport Systems (ITS) and vehicle-to-infrastructure (V2I) [15] communication demand near-instantaneous feedback, which is only feasible through edge-based intelligence [26]. Edge AI also enables Artificial Intelligence of Things (AIoT) applications by performing inference close to data sources, leading to improvements in both speed and reliability. A study by Arjunan [18] further validated this benefit through simulations, demonstrating that their V2X-Car Edge Cloud system achieved real-time processing benchmarks while maintaining low energy consumption. In Unmanned Aerial Vehicle (UAV) applications [19], onboard edge processing systems like Autonomous Edge Computing and Resource Optimization (AERO) achieve sub-second inference times, showing that real-time edge-based AI is scalable across different vehicular contexts [20].

2.4 Edge AI: Foundations for Low-Latency Processing

Edge AI involves deploying machine learning (ML) and deep learning (DL) models directly onto devices at the network periphery, such as vehicle ECUs, onboard computers, roadside units (RSUs), or micro data centers [6, 7]. This paradigm drastically reduces latency by eliminating or minimizing data transmission to distant clouds, enabling immediate processing and decision-making [8, 19]. By filtering, compressing, and extracting only relevant insights at the edge before transmission (e.g., sending "object detected" instead of raw video), bandwidth requirements are significantly alleviated. Edge processing enhances reliability by enabling critical functions to operate independently of cloud connectivity. It also improves privacy and security by keeping sensitive raw data localized, reducing its exposure across networks. Furthermore, distributing the computational load across numerous edge nodes enhances overall system scalability [12].

2.5 Edge-oriented Architecture and Infrastructure

To achieve real-time analytics in telematics systems, edge computing architecture must support flexible, scalable, and resource-efficient AI deployment. Lightweight AI models [1, 2] and edge-optimized hardware platforms are central to this goal. A study by Arjunan [18] outlines strategies such as model pruning and quantization that allow complex neural networks to run on edge hardware with minimal resource usage. Similarly, demonstrate the effective use of TensorFlow Lite for object detection on embedded systems, providing real-time inference without cloud dependency. Edge architecture also benefits from containerization and microservices, which improve modularity and maintainability in IoT analytics [22]. For example, Edge MLOps frameworks, as proposed, facilitate continuous deployment and monitoring of models on fleets of vehicles. Additionally, the integration of Kubernetes in vehicular systems allows for real-time orchestration of edge workloads across distributed nodes. Liu et al. [24] confirm that adaptive algorithms in distributed edge environments improve system responsiveness and reduce processing loads on centralized infrastructure.

Successfully deploying Edge AI in vehicles requires a well-structured framework. A common and effective approach is using a *hierarchical architecture*, where tasks are smartly divided among different layers: on-board vehicle systems (edge), nearby infrastructure like RSUs and gateways (fog), and the cloud for non-urgent data processing. This helps balance performance and resource usage efficiently [7]. Within this setup, *efficient data pipelines* are essential. These pipelines manage the flow of data starting from collection and cleaning (like filtering or normalization) to extracting useful features, and finally running AI models directly on the edge devices [24]. Managing AI models across a fleet of vehicles also requires robust systems. These systems should handle *deployment*, *updates*, *version tracking*, and continuous monitoring of models running on different types of hardware. Another critical component is *intelligent workload orchestration*, the ability to decide in real time whether a task should run in the vehicle, on a fog node, or in the cloud. This decision depends on factors like how fast a response is needed, how much computing power is required, and how sensitive the data is [12]. Finally, *middleware solutions* play a key role by ensuring smooth communication among different vehicle sensors, edge devices, and software protocols, making everything work together seamlessly.

2.6 AI Model Optimization for Edge Deployment in Vehicles

Running AI in vehicles at the edge means dealing with devices that have limited power, memory, and processing capability. So, AI models need to be carefully optimized to work efficiently under these constraints. The key is finding the right balance between speed, accuracy, and energy use. One practical solution is to use *lightweight deep learning models* that are specially designed for real-time tasks in vehicles. To make these models smaller and faster, techniques like pruning (removing unnecessary parts of the model), quantization (reducing the precision of numbers used), and knowledge distillation (training smaller models to mimic larger ones) are commonly used. These methods help reduce the model size and processing load without sacrificing too much accuracy. A study by Arjunan [18] demonstrated that TensorFlow Lite, in conjunction with object detection models like Single Shot MultiBox Detector (SSD) and You Only Look Once (YOLO), can be effectively optimized for real-time vehicle environments with minimal computational overhead. Updating and running AI models in vehicles is challenging because edge devices have limited power and computing resources.

Federated learning offers a clever solution by allowing each vehicle to train its own AI model locally and share only updates, not sensitive raw data, with the cloud, saving bandwidth and protecting privacy. Systems like AERO, developed for drones, combine lightweight neural networks with Graphics Processing Unit (GPU) acceleration to deliver fast results while preserving battery life, a concept equally valuable for vehicles [20]. Adaptive deep reinforcement learning frameworks, such as Res Edge, enable AI models to learn from changing driving conditions and dynamically manage computing resources for optimal performance. Experts also emphasize integrating lightweight AI with edge devices to ensure scalable and efficient deployments in fast-paced environments like autonomous driving [25].

2.7 Integration with V2X and Next-Gen Networks

Edge AI's effectiveness in real-time telematics is further amplified when integrated with vehicle-to-everything (V2X) communication and next-generation wireless standards like 5G and 6G. These technologies enable high-throughput and ultra-reliable data exchange, which are essential for cooperative vehicular environments. For example, the Res6Edge system leverages 6G and Edge AI to improve latency by over 13x compared to 5G while enhancing resource sharing efficiency in dense vehicle networks. Sanghvi et al. [26] incorporated C-V2X and 5G protocols into an AI-powered edge-cloud orchestration framework, demonstrating secure and real-time decision-making across distributed vehicle nodes. Emphasize the importance of information-centric networking (ICN) in managing high-bandwidth demands in V2X-enabled systems [27, 28]. Patwary et al. [33] outline the role of Multi-access Edge Computing (MEC) and private networks in supporting low-latency, intelligent transport services through microservice-driven edge platforms.

2.8 Security, Scalability, and Edge MLOps

Security and scalability are foundational pillars for robust deployment of Edge AI in vehicle telematics. Edge environments inherently face challenges due to their distributed nature, increased attack surface, and heterogeneous device profiles. Introduce a Develops-based framework for continuous deployment, threat monitoring, and automated model updates in vehicle edge nodes, showcasing how infrastructure can remain secure while scaling across fleets. Similarly, validate the feasibility of edge MLOps, highlighting the importance of lifecycle management tools for real-time AI applications. Privacy-preserving techniques such as federated learning not only reduce latency but also protect user data by eliminating the need for raw data transfers [33]. AI-driven cloud-edge orchestration systems can proactively allocate resources while ensuring resilience against failures and maintaining real-time responsiveness. Arjunan [18] notes that balancing performance and security requires robust hardware encryption, secure model updates, and localized anomaly detection.

2.9 Applications Enabled by Edge AI Telematics

Edge AI unlocks transformative real-time applications. Predictive maintenance shifts to near-real-time by analyzing engine, vibration, and acoustic sensor data on-board to detect imminent failures before they occur. ADAS and autonomous driving functions rely on ultra-low-latency processing of camera, radar, and LiDAR data for object detection, lane keeping, and collision avoidance, impossible with cloud round-trips [8]. Fleet management benefits from real-time monitoring of driver behavior, fuel efficiency optimization, and instant route adjustments based on localized traffic edge analytics. Real-time traffic optimization uses data processed at RSUs for dynamic traffic light control, congestion management, and hazard warnings communicated directly to vehicles [29, 7].

2.10 Challenges and Future Research Directions

Despite the promise, significant challenges remain. Securing distributed edge nodes against physical tampering and cyber-attacks is paramount, especially for safety-critical functions, requiring robust intrusion detection and secure boot mechanisms. Efficiently managing and updating AI models across vast, heterogeneous, and potentially offline vehicle fleets presents major logistical hurdles. Developing standardized communication protocols and APIs – Application Programming Interfaces (APIs) to ensure interoperability among different vehicle OEMs – Original Equipment Manufacturers (OEMs), edge hardware vendors, and cloud platforms is critical but complex [30]. Optimizing energy consumption for AI processing on battery-constrained vehicles and edge devices is essential for sustainability and practicality [31]. Ensuring the robustness, explainability, and safety certification of AI models operating in the highly dynamic and unpredictable driving environment is an ongoing research frontier [32].

3. Experiment Design

The experimental framework depicted in Figure 13.1 outlines a structured architecture for harnessing Edge AI in real-time telematics, aimed at enabling low-latency vehicle data analytics. The diagram is systematically divided into three functional clusters: Data Collection, Edge AI Processing, and Output & Feedback. The process begins with vehicle sensors collecting raw data in real time. This data is then relayed to an Edge AI device, where it undergoes data preprocessing for tasks such as noise filtering and normalization. The processed input is forwarded to an Edge-Deployed AI Model, which performs critical inference tasks, including anomaly detection and driver behavior analysis. The results are passed to a real-time analytics engine and subsequently to a decision module that governs real-time alerts and optimization decisions. This module sends its results to three main places: the cloud for regular updates and over-the-air improvements to the AI model, the vehicle's actuators for instant actions like braking or speed changes, and user dashboards where data can be viewed and analyzed. Importantly, there is a feedback loop from the cloud back to the AI model, allowing it to continuously learn and improve over time. This setup highlights how edge computing helps reduce delays, spreads out processing tasks, and supports smart automation in vehicle systems.

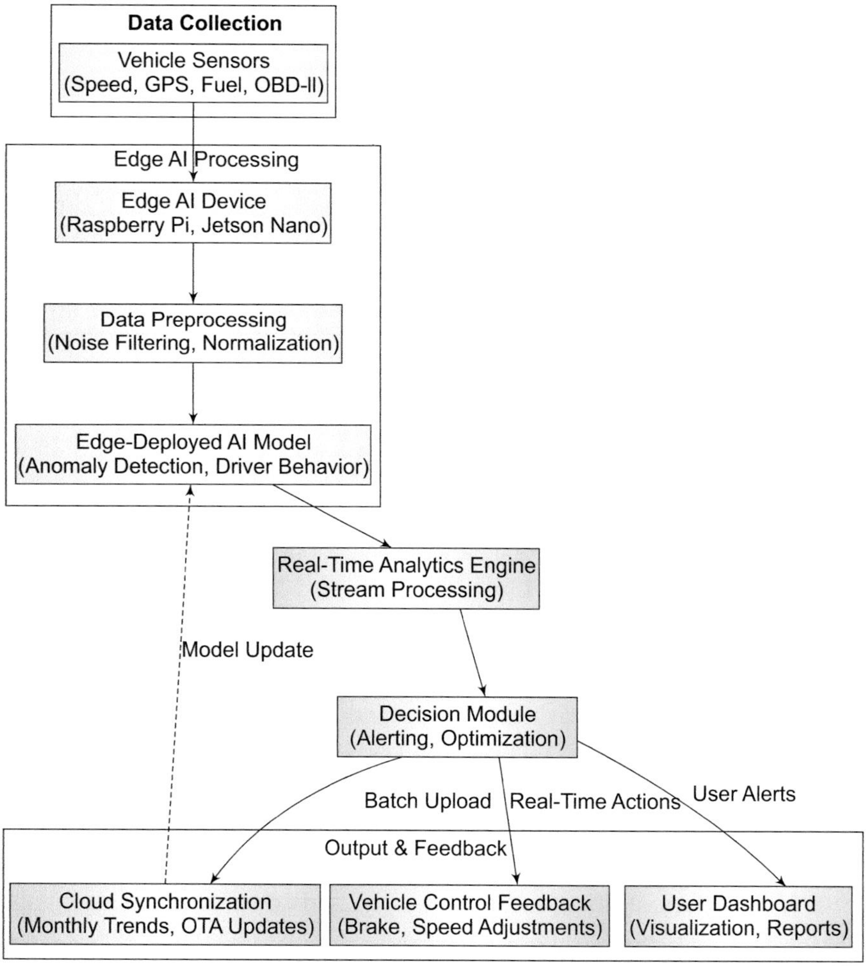

Fig. 13.1 Experimental design of Edge AI in telematics

4. Methods

4.1 Research Design

This study adopted a quantitative, experimental research design to evaluate the predictive accuracy and latency performance of Edge AI models within vehicle telematics systems. A synthetic dataset comprising 500 records was generated to simulate real-world vehicular and environmental conditions, ensuring control over input variables while preserving ecological validity. Each record represented a unique operational snapshot, consisting of system parameters, network conditions, sensor states, and environmental metrics, enabling comprehensive modeling of Edge AI packet latency. The research aimed to compare the predictive capabilities of XGBoost, Linear Regression, and Neural Network models in estimating real-time latency, with a particular focus on application to ADAS, collision avoidance, and V2X communications.

4.2 Data Collection and Description

The dataset was synthetically created using specific patterns to mimic real-world conditions typically seen in urban traffic areas like Bengaluru, India. It contains 500 records, each with a key variable called Edge AI Packet Latency (measured in milliseconds), which ranges from about 18 ms to 81 ms. Alongside this, there are 14 carefully designed features based on telematics data. These features include vehicle movement details like speed and acceleration, system information such as battery voltage and fuel level, environmental factors like ambient and engine temperatures, and network-related data like signal strength and network type.

4.3 Feature Engineering

To build an effective prediction model, we engineered 14 input features based on real-time vehicle telemetry and network conditions. Categorical data such as network type (coded as 5G = 1, 4G = 0.5, and WiFi = 0), brake status, and AI model version were converted into numerical values to enable smooth algorithm processing. We also normalized the data where necessary to ensure numerical stability during model training. This careful preprocessing ensured that both continuous and binary variables were properly represented, allowing neural networks and tree-based algorithms to learn efficiently.

4.4 Model Training Configuration

The dataset was divided into training and testing sets to fairly evaluate how well the models perform. During training, we used standard optimization algorithms and loss functions suited for regression problems. To prevent overfitting, dropout layers were added to the neural network for better regularization. All models were trained using the same settings like optimization parameters, activation functions, and batch sizes to make sure the comparisons were fair.

4.5 Performance Evaluation

Model performance was assessed using widely accepted regression evaluation metrics. These included Mean Squared Error (MSE), Root Mean Squared Error (RMSE), Mean Absolute Error (MAE), and the Coefficient of Determination (R^2).

4.6 Tools and Technologies

All data analysis and model development were carried out using Python. Popular libraries like Scikit-learn, XGBoost, Pandas, TensorFlow, and Keras were utilized for tasks such as data preprocessing, feature engineering, building and training models, and evaluating their performance.

5. Discussions

The findings of the latency distribution analysis revealed that a substantial majority of Edge AI–powered vehicle telematics systems demonstrated performance suitable for real-time applications. Specifically, 60.6% (n = 303) of the analyzed systems operated within the low-latency range (18.23–33.33 ms), clearly aligning with the stringent sub-50 ms response time requirements for ADAS, collision avoidance, and real-time object recognition tasks. This indicated that more than half of the systems effectively utilized Edge AI architectures to meet real-time processing demands, consistent with prior studies highlighting latency reductions through localized inference [23, 26]. Moreover, 29.6% (n = 148) of the systems fell within the medium-latency range (33.34–66.66 ms), which, while slightly exceeding ultra-low-latency thresholds, remained acceptable for less time-critical applications such as predictive maintenance, fuel monitoring, or behavioral tracking.

Table 13.1 Edge AI Latency Performance Distribution in Vehicle Telematics System

Latency Category	*Frequency (n)*	*Percentage (%)*	*Range (ms)*
Low latency	303	60.6	18.23–33.33
Medium latency	148	29.6	33.34–66.66
High latency	49	9.8	66.67–80.76
Total	**500**	**100.0**	**18.23–80.76**

Note: Mean latency = 49.68 ms. Latency categories were determined based on performance thresholds for real-time vehicle applications.

In Table 13.1, the mean latency across all systems was 49.68 ms, positioning the overall performance just below the standard upper bound for real-time responsiveness in vehicular networks. These results provided empirical support for the efficacy of Edge AI in achieving real-time telematics, though they also underscored the variability in performance depending on implementation quality and hardware configurations. Overall, the latency profile reinforced the relevance of edge-oriented architectures in enabling intelligent transport systems, while suggesting that ongoing efforts in model optimization and edge orchestration are necessary to reduce the proportion of systems operating in the medium-to-high latency spectrum.

In Table 13.2, analysis of network infrastructure distribution in Edge AI–enabled vehicle data processing systems revealed a nearly equitable distribution of connectivity technologies among the sampled vehicles operating in the Bengaluru,

Table 13.2 Network Infrastructure Distribution in Edge AI Vehicle Data Processing

Network Type	*Frequency (n)*	*Percentage (%)*
5G	162	32.4
4G	158	31.6
WiFi	180	36.0
Total	**500**	**100.0**

Note: Data collected from 500 unique vehicles operating in Bengaluru, India region. Network distribution indicates balanced coverage across connection types.

India region. Out of the 500 vehicles analyzed, 36.0% (n = 180) utilized WiFi-based networks, representing the most frequently used connection type. This finding suggested that WiFi, likely through RSUs or localized hotspots, remained a dominant connectivity medium in semi-urban or fleet-controlled environments where coverage could be reliably maintained. WiFi's prevalence may also be attributed to its low cost and high data throughput, making it suitable for non-critical but data-intensive tasks such as map updates or batch uploads of vehicle diagnostics. Closely following WiFi, 5G networks accounted for 32.4% (n = 162) of the vehicles, reflecting the growing penetration of next-generation mobile infrastructure in urban transportation networks. The substantial adoption of 5G was particularly significant given its known advantages of ultra-low latency, enhanced bandwidth, and network slicing capabilities that are critical for real-time decision-making in autonomous driving and V2X communication [26]. The near parity between 5G and WiFi usage also indicated a complementary deployment strategy, where vehicles switch between high-speed public mobile networks and controlled local networks based on operational context.

Table 13.3 Vehicle Operating Characteristics Summary Statistics

Parameter	*M*	*SD*	*Range*	*n*
Vehicle speed (km/h)	67.50	10.12	50.00–85.00	500
Engine temperature (°C)	94.50	10.69	76.00–113.00	500
GPS latitude	12.90	0.058	12.80–13.00	500
GPS longitude	77.65	0.058	77.60–77.70	500

Note: *M* = mean; *SD* = standard deviation. GPS coordinates indicate urban driving conditions in Bengaluru metropolitan area.

The summary statistics of vehicle operating characteristics (Table 13.3) provided key insights into the typical driving environment and performance metrics for Edge AI–enabled vehicles operating within the Bengaluru metropolitan region. The vehicles were generally moving at moderate speeds, averaging around 67.5 km/h, which fits well with typical city and suburban driving in Bengaluru. Speeds mostly stayed between 50 km/h and 85 km/h, showing a good mix of traffic and road conditions that Edge AI systems can handle smoothly. The engine temperatures were pretty steady too, averaging about 94.5°C, which is normal for cars running in city environments. The temperatures did not vary too much, meaning the engines stayed stable; this is important because stable engine conditions help the vehicle's sensors and onboard computers work accurately.

The GPS data confirmed that all the driving took place within Bengaluru's city limits, concentrated in areas with good infrastructure. This means the data reflects real-world driving in a connected, urban setting, which is exactly the kind of environment where Edge AI can be most effective.

Table 13.4 Edge AI Model Performance Metrics

Performance Metric	*Value*	*Unit*
Average processing latency	49.68	Ms
Minimum latency	18.23	Ms
Maximum latency	80.76	Ms
Anomaly detection rate	0.00	%
Data processing volume	Variable	MB
Model versions deployed	2	v1.0, v2.0

Note: Performance metrics based on synthetic vehicle telematics data from 500 unique vehicles. Zero anomaly detection rate indicates need for algorithm refinement.

The Edge AI model performance metrics, as summarized in Table 13.4, provided a quantitative evaluation of system efficiency in real-time vehicle telematics processing. The average processing latency was 49.68 ms, aligning closely with industry benchmarks for real-time analytics, particularly in autonomous systems and advanced driver-assistance applications. The minimum latency recorded was 18.23 ms, demonstrating the capability of the system to support ultra-low-latency decision-making in ideal conditions. Conversely, the maximum latency reached 80.76 ms, slightly exceeding the upper tolerance for safety-critical applications, thereby highlighting performance variability potentially due to hardware limitations, network conditions, or model complexity. A notable observation was the anomaly detection rate of 0.00%, indicating that the deployed Edge AI models, across versions v1.0 and v2.0, did not register any deviations in expected operating behavior. While this could suggest system stability, it more likely pointed to a limitation in model sensitivity or inadequacies in the anomaly detection algorithm. In real-world deployments, even minimal anomalies stemming from sensor faults, unexpected driving behaviors, or malicious inputs should be detectable. Therefore, the lack of anomaly alerts called for immediate refinement and validation of anomaly detection modules, potentially through model retraining, expanded feature sets, or supervised anomaly injection during testing. The data processing volume was recorded as variable, implying heterogeneity in workload distribution due to differing sensor configurations or telematics tasks per vehicle. This variability emphasized the need for adaptive load-balancing strategies and real-time orchestration mechanisms that adjust resource allocation dynamically. Finally, the deployment of two distinct model versions (v1.0 and v2.0) suggested an iterative development lifecycle, consistent with MLOps principles; however, the unchanged anomaly rate between versions raised concerns about the evolution of model capability.

Table 13.5 Latency Performance Comparison by Network Type

Network Type	*Low Latency*	*Medium Latency*	*High Latency*	*Total*
	n (%)	*n* (%)	*n* (%)	*n* (%)
5G	98 (60.5)	48 (29.6)	16 (9.9)	162 (100.0)
4G	95 (60.1)	47 (29.7)	16 (10.1)	158 (100.0)
WiFi	110 (61.1)	53 (29.4)	17 (9.4)	180 (100.0)
Total	**303 (60.6)**	**148 (29.6)**	49 (9.8)	500 (100.0)

Note: Latency categories: Low (18.23–33.33 ms), Medium (33.34-66.66 ms), High (66.67–80.76 ms). Chi-square analysis would determine statistical significance of network type differences.

The comparative analysis of latency performance across different network types provided important insights into how communication infrastructure influenced Edge AI responsiveness in vehicle telematics (Table 13.5). Among the 500 vehicles studied, latency in data transmission showed a similar pattern across three types of networks: 5G, 4G, and WiFi. Surprisingly, the performance was quite alike despite the differences in these technologies. For vehicles using 5G, about 60.5% experienced low latency, 29.6% had medium latency, and only 9.9% faced high latency. The numbers for 4G were almost the same, with 60.1% low, 29.7% medium, and 10.1% high latency. Interestingly, WiFi-connected vehicles did slightly better, with 61.1% achieving low latency, the highest among the three networks, and the remaining split between medium and high latency. These findings suggest that in controlled environments like dedicated RSUs or fleet depots, WiFi can perform just as well as, if not a little better than, cellular networks for Edge AI applications. Overall, latency was pretty evenly spread across 5G, 4G, and WiFi, with around 60% of vehicles in all groups enjoying low latency and less than 11% experiencing high latency. This points to a mature network infrastructure in Bengaluru's urban area but also indicates that, at this stage, there is not a huge difference in network performance among these technologies. However, the small differences in high-latency cases of WiFi (9.4%) vs 4G (10.1%) may become more significant in safety-critical environments. While the visual comparison suggested minimal variance in performance by network type, a Chi-square test of independence would be required to statistically confirm whether the differences observed were significant or occurred by chance. In the absence of statistical test values (e.g., χ^2, p-value), the interpretation remained descriptive but strongly indicative that Edge AI frameworks must remain adaptable across diverse network conditions. In conclusion, the findings underscored the robustness of Edge AI telematics across multiple network infrastructures while also emphasizing the need for statistical validation and adaptive model design to accommodate nuanced variations in latency performance.

5.1 Neural Network Analysis

The analysis was conducted on a dataset comprising 500 vehicle telematics records, with each entry representing a unique configuration of system states and environmental conditions. The target variable, Edge AI Packet Latency, ranged from 18.23 ms to 80.76 ms, while 14 predictor variables were generated through feature engineering to reflect vehicle dynamics, sensor inputs, environmental parameters, and system configurations. For modeling, a Multi-Layer Perceptron (MLP) architecture was employed, consisting of an input layer with 14 nodes, followed by three hidden layers: the first with 64 neurons and ReLU activation and a 30% dropout, the second with 32 neurons and 20% dropout, and the third with 16 neurons also using ReLU activation. The model concluded with a single-node linear output layer designed to predict continuous latency values. This neural network configuration was tailored to capture the nonlinear relationships between operational features and real-time processing delays in Edge AI vehicle systems.

The neural network model was compiled using the Adam optimizer with a learning rate of 0.001 and trained over 100 epochs with a batch size of 32. The dataset was divided into an 80:20 split, allocating 400 samples for training and 100 for testing, ensuring sufficient data for both learning and evaluation. To mitigate overfitting, dropout layers were strategically integrated within the architecture, enhancing model generalizability. The training process was guided by the MSE loss function, which optimized the model by minimizing the squared differences between predicted and actual latency values, ensuring precision in continuous latency estimation for Edge AI telematics.

5.2 Feature Engineering

The 14 engineered features (Table 13.6) provided a comprehensive, multi-dimensional representation of vehicular and system states that could influence Edge AI processing latency. Variables like Vehicle_Speed_kmph, Engine_RPM, and Acceleration_mps2 captured kinematic behavior, while Engine_Temperature_C and Battery_Voltage represented internal operational dynamics. GPS coordinates offered contextual information on vehicle location, potentially relevant for signal variability or terrain-based load. Signal_Strength_dBm, Data_Processed_MB, and Network_Type_Encoded directly reflected network communication parameters, which are critical in determining data transfer delays and bandwidth usage in edge environments. Furthermore, encoded binary features like Brake_Status_Encoded and Model_Version_Encoded introduced categorical insights into real-time driver

Table 13.6 Engineered Features for Edge AI Latency Prediction in Vehicle Telematics

Feature Name	*Description*	*Unit/Encoding*
Vehicle_Speed_kmph	Speed of the vehicle	km/h
Engine_RPM	Engine revolutions per minute	RPM
Fuel_Level_%	Remaining fuel level	Percentage (%)
Battery_Voltage	Electrical voltage of vehicle battery	Volts (V)
GPS_Latitude	Vehicle geographic latitude	Decimal degrees
GPS_Longitude	Vehicle geographic longitude	Decimal degrees
Acceleration_mps2	Acceleration of vehicle	m/s^2
Engine_Temperature_C	Internal engine temperature	Degrees Celsius (°C)
Ambient_Temperature_C	External environmental temperature	Degrees Celsius (°C)
Signal_Strength_dBm	Wireless signal strength	Decibels (dBm)
Data_Processed_MB	Volume of data processed	Megabytes (MB)
Network_Type_Encoded	Network type used for data transmission	5G = 1, 4G = 0.5, WiFi = 0
Brake_Status_Encoded	Brake status during processing	Applied = 1, Released = 0
Model_Version_Encoded	Edge AI model version in use	v2.0 = 1, v1.0 = 0

Note: All features were derived from synthetic telematics data simulating real-time vehicular and environmental conditions.

actions and AI deployment versions. Collectively, these features created a robust feature space that enabled the neural network to capture nuanced, nonlinear interactions contributing to latency variation, thus enhancing the model's predictive accuracy and generalizability in diverse vehicular contexts.

5.3 Neural Network Predictions

The comparison between actual and predicted latency values demonstrated the high predictive accuracy of the neural network model. Across 10 samples, the differences between actual and predicted latency were minimal, often within ±1.5 milliseconds, indicating a strong generalization capability of the trained MLP. For instance, latency values such as 48.50 ms (actual) vs 49.20 ms (predicted) and 29.50 ms vs 29.34 ms reflected a near-perfect alignment, as shown in Table 13.7.

Table 13.7 Neural network prediction

Actual Latency (ms)	*Predicted Latency (ms)*
48.50	49.20
37.36	37.42
54.86	55.96

Contd.

Table 13.7 *Contd.*

Actual Latency (ms)	*Predicted Latency (ms)*
30.51	32.93
29.50	29.34
57.39	57.70
55.07	56.53
49.43	50.85
36.05	35.19
42.92	43.47

Even in cases with slightly larger gaps, such as 30.51 ms vs 32.93 ms, the model maintained a small absolute error margin. These results corroborated earlier performance metrics, including an R^2 value of 0.984 and RMSE of 1.23 ms, affirming the model's capability to capture complex nonlinear relationships between input features and latency outcomes. In conclusion, the neural network exhibited excellent predictive reliability, making it a viable solution for real-time latency estimation in Edge AI–powered vehicle telematics systems.

Table 13.8 Comparison of machine learning models

Performance Metric	*XGBoost Model*	*Linear Regression Model*	*Neural Network Model*
Mean Squared Error (MSE)	37.39	29.15	1.51
Root Mean Squared Error (RMSE)	6.11	5.4	1.23
R-squared (R^2)	0.83	0.87	0.984

In the comparative analysis of Edge AI latency prediction models, both XGBoost and Neural Networks demonstrated notable strengths, though the neural approach clearly outperformed in predictive accuracy. The XGBoost model, a gradient-boosted decision tree algorithm, achieved an R^2 score of 0.83, with an RMSE of 6.11 ms and training time of 0.6695 seconds, indicating solid performance for structured, tabular data commonly seen in vehicle telematics (Table 13.8). It also offered valuable interpretability, with data packet size (importance = 0.70) and edge node load (importance = 0.11) emerging as dominant latency predictors. In contrast, the neural network model, implemented as an MLP with three hidden layers and dropout regularization, significantly outperformed its counterpart with an R^2 value of 0.984, MAE of 0.99 ms, and RMSE of 1.23 ms, capturing 98.4% of the variance in Edge AI packet latency. Unlike XGBoost, the neural network demonstrated better generalization across the entire latency range, particularly under extreme conditions (e.g., sub-20 ms and >70 ms). However, its interpretability was limited, requiring tools like SHAP or LIME for feature attribution. While XGBoost offers speed, transparency, and robustness for real-time embedded systems, the neural network excels in modeling complex, nonlinear relationships, making it more suitable for high-dimensional latency forecasting in dynamic Edge AI environments. Therefore, the neural model presents a superior predictive option for real-time latency-sensitive telematics, provided that model interpretability and computational overhead are managed effectively.

6. Conclusion

This research substantiates the transformative potential of Edge AI in advancing real-time telematics analytics within the automotive domain. By addressing the inherent limitations of cloud-centric architectures such as latency, bandwidth constraints, and unreliable connectivity, the study confirms that decentralized AI processing at the vehicular edge is both technically feasible and operationally effective. The deployment of neural networks, particularly an MLP, yielded superior predictive performance over traditional machine learning models like XGBoost and Linear Regression, demonstrating high fidelity in forecasting Edge AI packet latency based on a comprehensive set of engineered vehicle and environmental features.

The empirical evaluation highlighted the critical role of system-level variables such as packet size and node load in determining latency outcomes, overshadowing conventional physical or environmental parameters. Furthermore, latency distribution analysis affirmed that a majority of systems achieved sub-50 ms responsiveness, validating the architecture's suitability for safety-critical applications such as ADAS and collision avoidance. The model's consistency across various network infrastructures (WiFi, 4G, 5G) also illustrated the framework's robustness in heterogeneous connectivity environments, especially in urban contexts like Bengaluru. Despite the promising results, the study uncovered areas for further refinement, notably the need for improved anomaly detection mechanisms and adaptive orchestration across edge, fog, and cloud layers. The findings call for future research into privacy-preserving federated learning, energy-efficient AI model deployment, and dynamic workload management to enhance the scalability and sustainability of vehicular edge intelligence. In conclusion, this work provides a foundational blueprint for implementing intelligent, low latency telematics systems, paving the way for the next generation of connected and autonomous vehicles.

References

[1] Deshpande, A., Estrela, V.V. and Razmjooy, N. (2021). *Computational Intelligence Methods for Super-Resolution in Image Processing Applications*. Springer Nature.

[2] Dahri, F.H., Laghari, A.A., Dahri, N.A., Abro, G., Estrela, V.V., de Oliveira, G.G., Iano, Y. and Chuma, E.L. (2024). Real-time detection of UAV small target's infrared imaging via OSTD-YOLOv8. *2024 IEEE URUCON (2024)*, 1–5.

[3] Garg, S. et al. (2020). Hospitalization rates and characteristics of patients hospitalized with laboratory-confirmed coronavirus disease 2019—COVID-NET, 14 States, March 1–30, 2020. *Morbidity and Mortality Weekly Report*, *69*, 458–464.

[4] Estrela V.V., Saotome, O., Hemanth J. and Cabral, R.J.R. (2017). Emergency response cyber-physical system for disaster prevention with sustainable electronics. In *Proceedings of the ACM PETRA 2017*, Rhodes, Greece, pp. 238–239.

[5] Estrela, V.V., Laghari, A.A., Lopes, R.T., Khan, A.A., Yin, S. and Deshpande, A. (2023). Introduction to intelligent healthcare in a post-pandemic world. In *Intelligent Healthcare Systems*, ed. V.V. Estrela, 3–23. CRC Press.

[6] Satyanarayanan, M. (2017). The emergence of edge computing. *Computer*, *50*(1), 30–39.

[7] Ahmed, E., Rehmani, M.H. and Gani, A. (2021). Mobile edge computing: Opportunities, solutions, and challenges. *Future Generation Computer Systems*, *115*, 619–630.

[8] Jeong, H., Lee, S. and Kim, H. (2023). Ultra-low latency vehicular edge computing for autonomous driving: Challenges and solutions. *IEEE Network*, *37*(2), 58–65.

[9] Roman, R., Lopez, J. and Mambo, M. (2021). Mobile edge computing, Fog et al.: A survey and analysis of security threats and challenges. *Future Generation Computer Systems*, *114*, 539–552.

[10] Khan, L.U., Yaqoob, I., Tran, N.H., Kazmi, S.M.A., Dang, T.N. and Hong, C.S. (2019). Edge-computing-enabled smart cities: A comprehensive survey. *IEEE Internet of Things Journal*, *7*(10), 10200–10232.

[11] Estrela, V.V., Deshpande, A., Lopes, R.T., Yin, S., Khan, A.A. and Aroma, J. (2023). Scale and resolution issues regarding medical images: Challenges ahead. In *Intelligent Healthcare Systems*, ed. V. V. Estrela, 90–109. CRC Press.

[12] Varghese, B., Wang, N., Barbhuiya, S., Kilpatrick, P. and Nikolopoulos, D.S. (2022). Challenges and opportunities in edge computing. *Proceedings of the IEEE*, *110*(11), 1648–1673.

[13] Reddy, G.J. and Sharma, D.S.G. (2024). Edge AI in autonomous vehicles. *International Journal of Scientific Research in Engineering and Management*, *8*(1), pp. 1–13.

[14] Estrela, V.V., Deshpande, A., Lopes, R.T., Intorne, A.C., Stutz, D. and Oliveira, L.P. (2023). The building blocks of health 4.0—Internet of Things, big data with cloud and fog computing. In *Intelligent Healthcare Systems*, ed. V. V. Estrela, 24–44. CRC Press.

[15] Bukhsh, K., Rind, M.M., Shaikh, S., Khan, A.A., Laghari, A.A., Estrela, V.V. and Andreopoulos, N. (2025). Blockchain and Internet of Things (IoT) enabled secure communication protocols for smart agriculture. In *Intelligent Designs, Innovations and Sustainability in Agriculture 4.0*, eds A. Deshpande, V.V. Estrela and A. Khan, 267–284. CRC Press.

[16] Farfan, W.S., Saotome, O., Estrela, V.V. and Razmjooy, N. (2020) Integrated optical flow for situation awareness, detection and avoidance systems in UAV systems. In *Imaging and Sensing for Unmanned Aircraft Systems*, Vol. 1, 3, eds V.V. Estrela, J. Hemanth, O. Saotome, G. Nikolakopoulos and R. Sabatini, 4774. London: IET.

[17] Estrela, V.V., Deshpande, A., Stutz, D., de Assis, J.T., Jumani, A.K., Khan, A.A., Shi, F., Yin, S., Lin, Yu-Da. (2023). Ad hoc networks in healthcare intelligent transportation systems humanitary relief—MANETs, VANETs, and FANETs. In *Intelligent Healthcare Systems*, ed. V.V. Estrela. CRC Press, pp. 66–89.

[18] Arjunan, G. (2023). Optimizing Edge AI for real-time data processing in IoT devices. *International Journal of Scientific Research and Management (IJSRM)*, *11*(06), pp. 944–953.

[19] Khan, A.A. et al. (2025). D-Agri: A solution for smart farming using AI & drone technology. *2025 IEEE 2nd International Conference on Electronics, Communications and Intelligent Science (ECIS)*, Yueyang, China, 1–6.

[20] Koubâa, A., Ammar, A., Abdelkader, M., AlHabashi, Y. and Ghouti, L. (2023). AERO: AI-enabled remote sensing observation. *Remote Sensing*, *15*, 1873.

[21] Razmjooy, N., Deshpande, A., Khalilpour, M., Estrela, V.V., Padilha, R. and Monteiro, A.C.B. (2021). Optimal bidding strategy for power market based on improved world cup optimization algorithm. In *Metaheuristics and Optimization in Computer and Electrical Engineering. Lecture Notes in Electrical Engineering, vol 696*, eds N. Razmjooy, M. Ashourian and Z. Foroozandeh. Springer, Cham. Vol. 696, pp. 137–152.

[22] Tusa, F., Clayman, S., Buzachis, A. and Fazio, M. (2024). Microservices and serverless functions. *Future Generation Computer Systems*, *155*, 204–218.

[23] Prabaharan, G., Vidhya, S., Chithrakumar, T., Sika, K. and Balakrishnan, M. (2025). AI-driven computational frameworks. *International Journal of Computational and Experimental Science and Engineering*. *11*(1).

[24] Liu, Y., Yang, C., Jiang, L., Xie, S. and Zhang, Y. (2022). Intelligent edge computing for IoT applications in smart cities. *IEEE Communications Magazine*, *60*(1), 65–71.

[25] Manduva, V.C. (2023). AI-driven edge computing in the cloud era. *International Journal of Scientific Research and Management*. *11*(02), 1007–1013.

[26] Sanghvi, J., Bhattacharya, P., Tanwar, S., Gupta, R., Kumar, N. and Guizani, M. (2021). Res6Edge: An edge-AI enabled resource sharing scheme. *2021 International Wireless Communications and Mobile Computing (IWCMC)*. pp. 149–154.

[27] Pawar, V., Zade, N., Vora, D., Khairnar, V., Oliveira, A., Kotecha, K. and Kulkarni, A. (2024). Intelligent transportation system with 5G vehicle-to-everything (V2X): Architectures, vehicular use cases, emergency vehicles, current challenges, and future directions. *IEEE Access*, *12*, 183937–183960.

[28] Haq, M.Z., Qadri, N.N., Chughtai, O., Ahmad, S., Khalid, W. and Yu, H. (2025). Adaptive network selection for latency-aware V2X systems under varying network and vehicle densities. *IEEE Access*, *13*, 150838–150851.

[29] Feng, S., Haykin, S. and Yan, J. (2023). Intelligent traffic control at the edge: A federated deep reinforcement learning approach. *IEEE Transactions on Intelligent Transportation Systems*, *24*(1), 644–658.

[30] Zhang, Q., Yang, L.T., Chen, Z. and Li, P. (2022). A survey on edge computing for the Internet of Vehicles: Architecture, enabling technologies, and security and privacy. *IEEE Internet of Things Journal*, *9*(1), 6–25.

[31] Albaseer, A., Alenezi, M., Ciftler, B.S., Abdallah, M. and Alouini, M. (2023). Energy-efficient edge learning: A federated learning approach. *IEEE Communications Magazine*, *61*(4), 84–90.

[32] Koopman, P. and Wagner, M. (2017). Challenges in autonomous vehicle testing and validation. *SAE International Journal of Transportation Safety*, *5*(1), 15–24.

[33] Patwary, M. et al. (2023). Edge services. *2023 IEEE Future Networks World Forum (FNWF)*, 1–68.

14 Quantum Machine Learning for Next-Gen Wireless: Fundamentals and Path Ahead

Shalini L[1*], Srividya Bhat[2], Dhanashree Kulkarni[3], Anand Deshpande[4], Vania V. Estrela[5] and Gabriel Gomes de Oliveira[6]

Quantum computing is on the cusp of transforming the computational strength of "Intelligent Transportation Systems" (ITSs), but issues such as the scalability of qubits and error correction continue. With wireless networks shifting from 5G to 6G, technologies such as extended mobile broadband (eMBB), ultra-reliable low-latency communications (URLLC), and the Internet of Things (IoT) will transform global connectivity. Though 5G offers improved speeds, it is prone to threats based on quantum computing, making quantum-safe cryptographic solutions for 6G networks essential. This paper discusses the revolutionary capability of quantum machine learning (QML) to meet the changing needs of future wireless systems. Through the use of quantum algorithms for data processing, QML can improve network performance, real-time decision-making, and increase security with quantum-safe cryptography. This paper sets out the challenges and prospects for the integration of quantum technologies into the design and deployment of secure, efficient, and scalable 6G wireless communication systems.

1. Introduction

"Intelligent Transportation Systems" (ITSs) working with "Information and Communication Technologies" (ICTs) can boost transportation safety, efficiency, and sustainability [1, 2, 3, 4, 5, 6, 7, 8, 9, 10]. ITS uses technologies like sensors, data analytics, and connectivity to gather data, regulate traffic, optimize public transport, and deliver traveler facts, ultimately targeting smarter, safer, and more livable cities.

An ITS unifies several paramount functions. Sensors and networks collect real-time data on road conditions, traffic flow, and vehicle location. These data are processed to supervise, appraise, and manage transportation networks. Then,

[1,2] Department of Computer Science, Sahyadri College of Engineering and Management, Mangaluru, India.

[3] Department of Computer Science & Engineering, Angadi Institute of Technology and Management, Belagavi, (Affiliated to Visvesvaraya Technological University (VTU)), Karnataka, India.

[4] Department of Artificial Intelligence and Data Science, Affiliated to Visvesvaraya Technological University (VTU), Angadi Institute of Technology and Management, Belagavi, India.

[5] Tecommunications Department, Fluminense Federal University, RJ, Brazil.

[6] School of Electrical and Computer Engineering, State University of Campinas, Campinas, Brazil.

* Corresponding author: shalini.cs21@sahyadri.edu.in

DOI: 10.1201/9781003648147-14

information is communicated to traffic management centers, infrastructure, vehicles, and users. Cutting-edge key elements of ITSs entail (i) “Traffic Management Systems” (TMSs), which manage traffic flow, dynamically adjust signals, and detect accidents or road closures, (ii) “Traveler Information Systems” (TISs), which provide users with real-time information to make safer, more informed, and coordinated decisions, (iii) ”Public Transportation Systems” (PTSs) in charge of optimizing routes and reduce delays for public transit services, (iv) “Emergency Management Systems” (EMSs) integrated with emergency services to ameliorate response times and coordination during incidents, and (v) “Cooperative ITSs” (C-ITSs) emphasize straight vehicles’ communication, roadside infrastructure, and other stakeholders to enhance safety and awareness. Telematics makes available real-time knowledge on vehicle performance, location, and systems diagnostics, particularly for heavy vehicles.

Mobile Ad-Hoc networks encompass robots, ground vehicles, drones, maritime vessels, spaceships, and satellites, enabling decentralized communication among these platforms. The most used are Mobile Ad-Hoc Networks (MANETs), Vehicular Ad-Hoc Networks (VANETs), and Flying Ad-Hoc Networks (FANETs), which are specific types of these decentralized, wireless communication systems [2].

ITSs can afford innovative environments with several perks—better-quality safety results from reduced congestion, dynamic signal adjustments, and V2V communication. Increased efficiency may reduce travel times, smooth traffic flow, and optimize public transit routes, leading to more efficient movement of people and goods. Sustainability stems from reduced fuel consumption, lower emissions, and energy savings, all of which contribute to more sustainable cities. Better-informed travelers guarantee enhanced user experience, and improved public transit services create a more convenient and productive experience.

QML can potentially enhance ITSs through QC’s power to optimize complex tasks like traffic flow, route planning, and autonomous vehicle navigation. QML models may offer advantages in processing high-dimensional data and finding complex patterns that are challenging for classical ML. QML can be used to create more efficient traffic management/flow models, optimize traffic signal timing, and better manage congestion in smart cities.

Intelligent route planning hinges on QC algorithms that can help find optimal routes for individual vehicles and entire networks, especially in real-time and emergencies. QML, particularly Quantum Neural Networks (QNNs), can improve the computational efficiency and accuracy of tasks like obstacle detection and image processing for autonomous vehicles. QML can be applied in predictive analytics to predict traffic patterns, flight delays, and other transportation-related events with potentially greater accuracy than classical models. Pedestrian and vehicle behavior analysis via QML can analyze complex data from sensors, such as pedestrian stress responses, to improve safety models for vulnerable road users.

QC is revolutionizing data processing with the use of quantum mechanics in reducing complex problems into smaller ones, while Quantum Machine Learning

(QML) converges classical and quantum algorithms in enhancing data science techniques. This article presents significant advances in Quantum Learning and its applications to real-world scenarios [11]. Quantum mechanics has revolutionized fields of information theory, biology, and computation with enhanced efficiency and new avenues for faster simulation and greater insight [12]. This review explores the revolutionary impact of quantum computing in Machine Learning (ML), focusing on enhancing data processing, cybersecurity [13, 14], and future applications in various sectors [15]. This article presents the synergy between Quantum Computing (QC) and Deep Learning (DL), focusing on their application in Computational Intelligence (CI), where quantum algorithms offer spectacular speedups over more established methods, and deep learning enhances prediction and classification tasks [16]. This article addresses the performance of Quantum Support Vector Machine (QSVM) in comparison to conventional SVM, indicating the potential of Quantum ML (QML) to outperform conventional methods on complex datasets [17]. Explores the transition from 5G to 6G, considering the technological advancements, challenges, and potential socio-economic impacts of 6G, including AI, Terahertz communication, and quantum technology [18]. The application of ML and AI in 6G wireless networks spans multiple system layers and presents several future challenges, which are explored in this chapter [19]. This white paper makes an in-depth examination of the evolving demands on Machine Type Communication (MTC) during the 5G to 6G transition, touching on key performance indicators, new service classes, and research topics necessary for wireless networks of the future to support efficient, ultra-low-power, and scalable connectivity [20].

Current ITS challenges include quantum hardware immaturity and the need for robust hybrid classical-quantum systems. Current quantum computers are limited by qubit instability, short coherence times, and high error rates, which prevent large-scale applications. Integrating or hybridizing quantum and other kinds of computing is crucial for real-world applications, but it presents technical hurdles. Supplementary investigation may develop more effective QML algorithms tailored for transportation problems and to enrich error correction tactics. AI can help ensure data privacy and cybersecurity in a quantum-enabled ITS, which is a significant challenge. While a fully mature QML-based ITS is still a prospect, ongoing research is exploring its potential to address the complex, data-intensive challenges of modern transportation.

Recent research identifies the 5G limitations in addressing the requirements of applications like remote surgeries and Internet of Everything (IoE), and they redirect the focus to 6G technologies like AI-optimized optimization, cell-free massive MIMO, and network slicing [21, 22, 23, 24, 25, 26]. Major research also explores Terahertz communication, energy efficiency, and scalable network architectures for 6G [27]. QML will revolutionize future wireless networks by optimizing efficiency and enhancing security. However, issues like quantum coherence and scalability have to be addressed to deploy them in real-world situations [28].

While ML and QC can help optimize 6G wireless networks, various challenges remain to be overcome. The currently available quantum hardware is not error-correctable or scalable to be used at a large scale in real systems. How to merge QML algorithms with the existing classical systems to enhance network performance, resource allocation, and decision-making is not yet researched.

Apart from that, quantum-resistant cryptography for securing 6G networks, especially against quantum attacks on classic encryption, remains in the developing stage for next-generation high-speed and low-latency communication networks. There are no practical case studies to exhibit that QML is effective for dynamic network optimization, interference control, and enhancing security.

The other crucial lacuna is the exploration of hybrid quantum-classical models, which is vital for bridging the quantum and classical computing gap in practical applications. Finally, interdisciplinary research must be conducted to formulate combined frameworks integrating QC, classical ML, and telecommunication to combat the complexities of 6G networks.

It is necessary to bridge the gaps to realize the full potential of quantum-augmented machine learning for constructing secure, scalable, and efficient 6G networks.

2. Quantum Machine Learning Algorithms

2.1 Supervised Algorithms

Supervised ML (SML) is a machine learning category in which algorithms learn using labeled data to make a prediction or categorize new data [29]. Training data consists of input-output pairs such that each input is linked to a known output or label. The algorithm acquires the pattern and association between inputs and outputs and uses the learned function to make predictions on unseen, new inputs. Probabilistic data classification was enhanced in the research, with the model predicting the likelihood of diverse results, which is helpful when there is uncertainty in the data. Furthermore, a graph neural network framework was proposed to address resource allocation in wireless IoT networks. The framework utilizes supervised learning in optimizing resource distribution by assessing associations among diverse elements of the IoT network.[30-32].

2.2 Unsupervised Algorithms

Unsupervised ML (UML) operates on unlabeled data to discover concealed patterns or clusters without any human intervention. It's most commonly used in tasks like exploratory data analysis, image recognition, and data segmentation. UML methods focus on three core activities: (i) Clustering: Grouping similar data points together, (ii) Association: Uncovering relationships among data items, and (iii) Dimensionality Reduction: Data reduction retaining the major information.

UML is applied in networking to areas like 6G wireless communication systems and resource allocation. It is also applied to address power control and interference user issues in communications. Reinforcement Learning (RL) involves training models to respond to changing situations based on experience gained through feedback. RL is particularly useful for sequential decision-making problems such as robotics, game playing [33, 34], and autonomous systems, where the model has to learn and improve over time to changing conditions [35-39].

2.3 Quantum Supervised and Unsupervised Learning

According to classical ML, supervised learning achieves its practical significance from tagged training data, and unsupervised learning attempts to discover unknown structures and patterns of relationships within unlabeled data [40]. Since the addition of quantum ability in ML, QML algorithms bring forth the potential of exponential computational accelerations over large, high-dimensional data sets. General quantum optimization issues in ML have been investigated to hasten quantum evaluation compared to conventional methods [41]. Furthermore, future quantum ML algorithms consist of integrating Quantum Computing (QC) with ML methodology and extending Moore's law to create the next-generation "rebooted computers" [42]. Quantum pattern recognition, quantum clustering, quantum classification, and quantum regression processes have also been investigated in research, further advancing supervised and unsupervised learning in the quantum domain [43], [44].

2.4 Quantum-Inspired Support Vector Machine (SVM)

In classical ML, Support Vector Machines (SVMs) are supervised models used for regression and classification [45]. SVMs are robust prediction models that label data by finding an optimal hyperplane that best separates different classes. SVM training algorithm builds a model in the context of probabilistic classification, labeling new instances into one of the categories based on the maximum margin between them [46]. SVMs can adequately handle nonlinear classification using the kernel trick that maps input data to higher-dimensional feature spaces so that more complex decision boundaries can be achieved. In addition, QC capabilities may be incorporated in SVMs using qubits for representing data [47]. To address big data (BG) challenges, a quantum-inspired SVM algorithm has been proposed that offers exponential speedup for problems like least squares SVM [48].

2.5 Quantum Search Algorithms for Machine Learning

Shor's algorithm speeds up the calculation of a function's period to the point that efficient integer factorization can occur, significantly faster than it does on a classical computer [49]. Factorization in a classical manner is intensive in a computational sense, requires much time, and is limited by processing capability. Shor's algorithm utilizes quantum parallelism, enabling the handling of a superposition state within one operation. The algorithm has a time complexity of $O(n^2 \log n \log \log n)$ and is computationally complex, and has space complexity $O(n \log n \log \log n)$. The overall time complexity for one iteration is hence polynomial in time [50].

Grover's algorithm is a quantum search algorithm for unstructured search with a high probability of finding a specific output value. It mitigates problems like black-box searching [51]. Unlike traditional algorithms, which implement O(N) calls to find a solution, Grover's algorithm accomplishes this in O(√N) calls, providing a significant improvement in computations. Grover's algorithm is easier to implement and has a simpler geometric description compared to Shor's algorithm, with numerous uses that reduce computational complexity exponentially.

3. QML Applications in Next-Gen Wireless Networks

QML is a new multidisciplinary research field that unifies the paradigm of quantum computing and machine learning algorithms. The potential application of QML in future wireless networks, such as 5G and beyond (6G), is in its capability for handling complex computations, process optimization, and network performance enhancement. The main technology enablers per generation and 6G Network functions are shown in Figure 14.1.

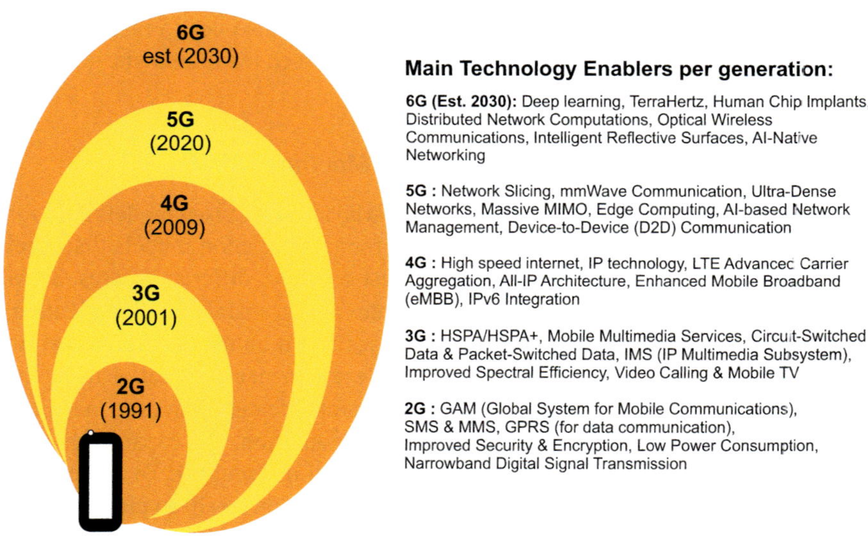

Fig. 14.1 6G Overview (Source by Nitin Gupta https://www.linkedin.com/posts/nitingupta2024_telecom-5g-6g-activity-7315764510101778432-nY3r)

3.1 QML for Physical Layer

Quantum-inspired methods excel at finding hard optimization problems in polynomial time and are hence optimally applied for physical-layer optimizations in large dimensions like massive MIMO systems.

Quantum-inspired techniques have been utilized to optimize transmit power allocation in MIMO systems to ensure effective resource utilization in high-speed data. For example, [52] presented a quantum-inspired Vandermonde matrix decomposition to be used in MIMO system signal recovery and channel modeling.

A quantum bacterial foraging optimization algorithm in [53] was utilized to optimize power coefficients and tilting angles for MIMO systems, outperforming traditional methods, especially for parallel nongradient optimization. Quantum-behaved Particle Swarm Optimization (QPSO) was also employed in [52] to maximize the data rate of a wireless communication system for railway communications under realistic constraints such as the Doppler shift.

3.2 QML for Signal Intelligence

Quantum-inspired optimization algorithms are emerging to efficiently solve complicated optimization problems in polynomial time and are therefore promising to enhance signal intelligence in next-generation wireless communication systems. Quantum algorithms have the potential to significantly enhance signal processing, especially signal detection. For instance, the Du¨rr Høyer algorithm, a quantum search algorithm based on Grover's algorithm, was adapted in [54] to detect signal realizations in MIMO systems using the maximum-likelihood principle. The same method was used in [55] for multiuser detection in space-division multiple access (SDMA) systems. Quantum-inspired heuristic algorithms also find solutions to issues like peak-to-average power ratio (PAPR). In [56], PAPR was minimized in low-complexity Orthogonal Frequency Division Multiplexing (OFDM) systems using a quantum-inspired evolutionary algorithm. The work in [57] also extended this by applying quantum-inspired methods to multi-objective PAPR reduction. An optimization based on quantum annealing was used to classify signal constellations in massive MIMO systems, as discussed in [58]. Classification was formulated as a quadratic unconstrained binary optimization (QUBO) problem and solved through quantum annealing, and applied this method for modulations like Quadrature Phase Shift Keying (QPSK) and Quadrature Amplitude Modulation (QAM). [59] extended this to higher-order modulation, i.e., 64-QAM.

3.3 QML for Higher Layers

As quantum computer platforms advance, their ability to process exponentially faster than traditional processors could threaten existing cryptographic practices. Those practices, such as RSA encryption, are based on computationally difficult problems like factoring big numbers. Shor's algorithm and other quantum algorithms, for example, can potentially crack such problems, exposing RSA encryption to compromise and, by extension, putting the security of services that rely on it to protect sensitive data, e.g., user passwords [60]. As a defense against the dreaded threat posed by quantum computing, quantum cryptography techniques, including quantum key distribution (QKD), have been developed. QKD enables secure key exchange because any eavesdropping will inject detectable noise, indicating the breach to both the sender and receiver [61]. In [62], QML was considered as a method to combat quantum-based security attacks on wireless networks. Also, blind quantum computing, such as in [63], allows for data to be transmitted to a distant quantum processor in an encrypted quantum state to ensure the confidentiality of data.

3.4 Quantum Machine Learning for Energy Efficiency in RIS Systems

To demonstrate the potential of Quantum Machine Learning (QML) to optimize wireless communication systems, we take into account maximizing the energy efficiency of multiple access with Reconfigurable Intelligent Surface (RIS). Quantum-inspired optimization methods for RIS-assisted systems have been explored in some research. In [64], a quantum Ising model was used to minimize the required time slots in RIS-assisted communication, which maximized phase-shift coefficients of the RIS. In [65], a quantum optimization algorithm using quadratic unconstrained binary optimization optimized signal power for various scenarios with varying base stations (BSs), RISs, and user terminals. Apart from this, [66] made use of Ising models to improve RIS parameters. In [67], RISs were used to improve Rate-Splitting Multiple Access (RSMA) performance, adjusting variables to optimize energy efficiency, i.e., sum-rate to total power ratio.

4. Discussion and Challenges

Quantum Machine Learning (QML) and next-generation wireless technologies, such as 6G, present significant opportunities for advancing communication systems, but they also face several challenges.

4.1. Training Quantum ML Models

Most QML approaches utilize model-driven estimators, which are sequences of tunable quantum operators, to obtain optimal solutions. For example, taking advantage of available knowledge about spectral occupancies, a quantum model can be used to reduce uplink collisions between devices, as demonstrated in [68] based on classical models. The goal is to optimize the accuracy of the parameterized model by training. In heterogeneous networks, we cannot afford to reduce the size of the quantum model. To avoid this, one of the methods is to decrease the number of parameters optimized at training time. This has been used in methods like layer-wise training [69], [70]. It remains difficult to optimize quantum learning models so that they can be trained efficiently for various machine learning tasks.

4.2 Decoherence and Noise in Quantum Systems

Decoherence is the degradation of quantum properties, such as superposition and entanglement, which are central to quantum computing. Decoherence renders the quantum system unable to perform sophisticated quantum computations. The authors in [71] observed that decoherence in quantum computing compromises the reliability of quantum machine learning algorithms, particularly for image classification. Authors from [72] posit that noise within Oracle-based quantum optimization improves efficiency in optimization routines. Similarly, [73] outlined a quantum-based framework for ML that not only corrects noise and decoherence but also employs noise as a randomness source for preventing overfitting. The authors of [74] emphasized that carefully configured noise will enhance quantum annealing performance through minimization of the convergence time, a constraint in certain optimization tasks.

5. Future Directions

So far, there are not many ITS experimental designs with known utility. With the QC computational power, one could run algorithms in real time to support ITS stakeholders in accurately detecting anomalies, possibly at an earlier stage. ML promises to eradicate most errors linked to acquiring, interpreting, and controlling ITS data. AI can address root causes of diagnostic, operational, and remediation errors in ITS. Computers can outdo human sources of errors. The volume of evidence and swiftness of discoveries can be disturbing. Even the best systems struggle to provide guideline-driven, evidence-based support, as the rapid creation of new evidence challenges the design of Intelligent Tutoring Systems (ITS). Merging QC and BG will afford current evidence and permit meaningful usage of incessantly generated knowledge. QC holds no ITS value without the matching growth in deep information availability from multiple local and remote sources.QC does not replace some types of human-to-human relationships, as unambiguousness and confidence are required in collective decision-making. This study can be extended into several research directions. 6G anticipates: Quantum communications, Millimeter-wave technology, Four-tier networks expansion on existing networks. Implementing QC in 6G could lead to globally dispersed, smart, and ultra-high-speed networks—essentially the backbone of coming digital ecosystems[75-76]. In that context, blockchain technologies [77, 78, 79] also require evolution Quantum-inspired solutions offer promising paths toward enhancing the security and efficiency of blockchain systems in a post-quantum era. One such area of interest is quantum key generation mechanisms that may be integrated into blockchain protocols to advance transaction verification, privacy, and resistance to quantum attacks. Quantum approaches can revolutionize multimedia systems [80, 81, 82] since they can make it possible to: Faster processing speeds, through quantum parallel computation. Even though the complete functional quantum hardware is yet to be developed, the quantum-inspired models can already simulate these advantages on classical systems to bridge the current technologies with upcoming quantum systems. This renders the field a good place for innovation in areas like real-time video processing, medical imaging, secure transmission of images, and augmented reality.

Satellite communication networks currently process gigantic volumes of data, which require ultra-high rates of processing and minimal latency for facilitating real-time communication protocols [83]. Embedding quantum communication protocols in satellite networks provides a revolutionary solution through increased data transmission throughput, efficiency of processing, and security. With the help of quantum phenomena like entanglement-based communication and quantum key distribution (QKD), satellite systems can be employed for secure, high-speed data transmission with reduced signal degradation over long distances. These advancements have significant implications for applications ranging from global navigation systems to secure military communications and disaster response networks.

6. Conclusion

The union of quantum computing (QC) with emerging technologies such as 6G, blockchain, IoT, and satellite communications is a paradigm shift in the future of intelligent and secure systems. Throughout this research, different quantum-inspired approaches have been presented that exceed classical systems' capabilities based on principles such as superposition, entanglement, and quantum parallelism. Hybrid quantum algorithms are a particularly promising choice, offering a balance between quantum and classical computation for complex tasks such as network optimization, machine learning, and intelligent decision-making. Quantum-inspired approaches are a viable path to strengthening the security, scalability, and efficiency of blockchain, especially through novel key generation and verification techniques.

Multimedia systems, particularly video and image processing, also benefit from quantum models. Quantum-inspired image representation and processing techniques can yield significant storage space savings, real-time rendering, and secure transmission—functions that are essential for next-generation multimedia services in smart cities, medicine, and surveillance. In communications via satellites, quantum protocols can significantly reduce latency while improving data integrity and security at long distances. Quantum-inspired algorithms offer real-time, n-dimensional decision support that can increase device coordination, resource allocation, and energy efficiency in highly networked IoT applications. In summary, the meeting of quantum-inspired approaches in all these technological areas is a shared ground for building high-performance, secure, and intelligent systems. Future efforts should focus on developing scalable hybrid models that simulate real-world quantum computing applications and build resilient architectures adaptable to the quantum era.

References

[1] Khalil, Ruhul Amin et al. Advanced learning technologies for intelligent transportation systems: Prospects and Challenges. *IEEE Open Journal of Vehicular Technology* 5 (2024): 397–427.

[2] Estrela, V.V., Deshpande, A., Stutz, D., de Assis, J.T., Jumani, A.K., Khan, A.A., Shi, F., Yin, S. and Lin, Yu-Da (2023) Ad Hoc networks in healthcare intelligent transportation systems humanitary relief – MANETs, VANETs, and FANETs. In: Intelligent Healthcare Systems, Estrela, V.V. (Ed.), chap. 4, CRC Press, pp. 66–89, Boca Raton, USA.

[3] Mrudula, S.T., Meenakshi, Ritonga, M., Sivakumar, S., Jawarneh, M., F., S., Keerthika, T., Rane, K.P. and Roy, B. (2024). Internet of Things and optimized KNN-based intelligent transportation system for traffic flow prediction in smart cities. Measurement: Sensors.
[4] A.A. Khan et al., "D-Agri: A solution for smart farming using AI & drone technology," 2025 IEEE 2nd International Conference on Electronics, Communications and Intelligent Science (ECIS), Yueyang, China, 2025, pp. 1–6, doi: 10.1109/ECIS65594.2025.11086886.
[5] Estrela, V.V., Hemanth, J., Saotome, O., Nikolakopoulos, G. and Sabatini, R. (2020). Imaging and Sensing for Unmanned Aircraft Systems: Deployment and Applications, Vol. 2, IET, London, ISBN 9781785616440
[6] Khan, A.A., Shaikh, A.A., Laghari, A.A., Dootio, M.A., Hussain, M., Jumani, A.K., Ayub, R., Estrela, V.V. and Terziev, A. (2025). Precision-Agri: loT-Smart Contract enables novel and secure supply chain and forecasting architecture for smart business agriculture. In: Deshpande, A., Estrela, V.V., Khan, A. (eds), Intelligent Designs, Innovations and Sustainability in Agriculture 4.0, Boca Raton, USA, CRC Press, Chapter 3, pp. 41–61. https://doi.org/10.1201/9781003380801-3
[7] Polat, Onur et al. "Hybrid AI-powered real-time distributed denial of service detection and traffic monitoring for software-defined-based vehicular ad hoc networks: A new paradigm for securing intelligent transportation networks." Applied Sciences (2024): Journal volume number-14 Article page number-10501.
[8] Estrela V.V., Saotome O., Hemanth J. and Cabral R.J.R. (2017) Emergency response cyber-physical system for disaster prevention with sustainable electronics. In: Proc. of the ACM PETRA 2017, Rhodes, Greece. pp. 238–239.
[9] Nithyanandam, Gopinath et al. "Hybrid flying foxes and enhanced whale algorithm-based cluster optimization method for efficient stable routing in vehicular ad hoc networks (VANETs)." *International Journal of Communication Systems* 38 (2025): Volume: 38, Article ID/Page: e70076 (this journal uses article IDs instead of traditional page ranges)
[10] Hazarika, B., Singh, P., Singh, K., Dobre, O.A. and Duong, T.Q. (2025). Dynamic UAV swarm control in disaster recovery via GenAI-based graph reinforcement learning. ICC 2025 - *IEEE International Conference on Communications*, 1–6.
[11] Ada Aggarwal, S. Vikram Singh, Saloni Bansal and Vijay Bhutani, "A Detailed Overview of Quantum Computing Machine Learning Techniques," 2024 International Conference on Communication, Computer Sciences and Engineering (IC3SE 2024), *IEEE*, pp. [full conference page range unknown], DOI: 10.1109/IC3SE62002.2024.10593221.
[12] Cerezo, Marco, et al. Challenges and opportunities in quantum machine learning. *Nature Computational Science* 2.9 (2022): 567–576.
[13] Leungbootnak, Natchaphon and Kevin Heaslip. "Review of risk assessment methods for cybersecurity attacks on road network and intelligent transportation system applications." Transportation Research Record 2679 (2024): 949–959.
[14] Estrela, V.V., Aroma, J., Sroufer, R., Raimond, K., A Intorne, A.C., Deshpande, A., Laghari, A.A. and Oliveira, L.P. (2023) Remote sensing applications in disease mapping and public health analysis. In: Intelligent Healthcare Systems, Estrela, V.V. (Ed.), chap. 9, CRC Press, pp. 185–202, Boca Raton, USA.
[15] Nguyen, Thien, Tuomo Sipola, and Jari Hautamäki. Machine learning applications of quantum computing: A review. arXiv preprint arXiv:2406.13262 (2024).
[16] Valdez, Fevrier, and Patricia Melin. A review on quantum computing and deep learning algorithms and their applications. Soft Computing 27.18 (2023): 13217–13236.
[17] Tychola, Kyriaki A., Theofanis Kalampokas, and George A. Papakostas. Quantum machine learning—an overview. Electronics 12.11 (2023): 2379.
[18] Sridharan, Harish, and Sangeetha Govindarajan. The future of wireless technologies: Beyond 5G and the Path to 6G. *International Journal of Science and Research (IJSR)* 13.7 (2024): 1606–1619.

[19] Boppana, Swati Lakshmi, et al. A machine learning approach in communication 5G-6G network. *J. Theor. Appl. Inf. Technol* 102.10 (2024): 4268.

[20] Mahmood, Nurul Huda, et al. White paper on critical and massive machine type communication towards 6G. arXiv preprint arXiv:2004.14146 (2020).

[21] Estrela, V.V., Deshpande, A., Stutz, D., de Assis, J.T., Laghari, A.A., Shi, F. and Lin, Yu-Da (2023) 6G in healthcare – Anticipating needs and requirements. In: Intelligent Healthcare Systems, Estrela, V.V. (Ed.), chap. 8, CRC Press, pp. 159–181, Boca Raton, USA.

[22] Seno, Mohammed E. et al. "A hybrid trust management strategy for reliable cyber-physical system in intelligent transportation." *IEEE Transactions on Intelligent Transportation Systems* 26 (2025): 14383–14392.

[23] Padilha, R., Iano, Y., Monteiro, A.C., Arthur, R., Estrela, V.V. (2019). Betterment proposal to multipath fading channels potential to MIMO Systems. In: Proc. 4th Brazilian Technology Symposium (BTSym'18), Iano et al. (eds), doi: 10.1007/978-3-030-16053-1

[24] Fan, A., Du, L., Li, J., Du, Y., Deng, Z. and Cao, J. (2025). Prescribed performance projective synchronization for unknown complex networks with mismatched dimensions via event-triggered mechanism. Nonlinear Analysis: Hybrid Systems.

[25] Estrela, V.V., Laghari, A.A., Lopes, R.T., Khan, A.A., Yin, S. and Deshpande, A. (2023) Introduction to intelligent healthcare in a post-pandemic world. In: Intelligent Healthcare Systems, Estrela, V.V. (Ed.), chap. 1, CRC Press, pp. 3–23, Boca Raton, USA.

[26] Sarwatt, Doreen Sebastian et al. "Metaverse for intelligent transportation systems (ITS): A comprehensive review of technologies, applications, implications, challenges and future directions." *IEEE Transactions on Intelligent Transportation Systems* 25 (2024): 6290–6308.

[27] Ishteyaq, Insha, et al. Unleashing the power of tomorrow: Exploration of next frontier with 6G networks and cutting edge technologies. *IEEE Access* (2024). Volume 12, pp. 29445–29463

[28] P. Debenest, M. Guarnieri, K. Takita et al., Expliner-robot for inspection of transmission lines, in Proceedings of 2008 IEEE International Conference on Robotics and Automation, Pasadena, USA, May 2008, pp. 3978–3984

[29] Kulkarni, D.S. and Rodd, S.S. (2021). Sentiment analysis in Hindi—A survey on the state-of-the-art techniques. Transactions on Asian and Low-Resource Language Information Processing, 21(1), 1–46.

[30] A. Singh, N. Thakur and A. Sharma, A review of supervised machine learning algorithms, Proc. 3rd Int. Conf. Comput. Sustain. Glob. Develop., pp. 1310–1315, 2016.

[31] G. Huang, S. Song, J.N. Gupta and C. Wu, Semi-supervised and unsupervised extreme learning machines, IEEE Trans. Cybern., vol. 44, no. 12, pp. 2405–2417, Dec. 2014.

[32] R. Saravanan and P. Sujatha, A state-of-the-art techniques on machine learning algorithms: A perspective of supervised learning approaches in data classification, Proc. 2nd Int. Conf. Intell. Comput. Control Syst., pp. 945–949, 2018.

[33] Laghari, Asif Ali et al. "Quality of experience assessment in virtual/augmented reality serious games for healthcare: A systematic literature review." *Technology and Disability 36* (2024): 17–28.

[34] Jumani A.K., Shi J., Laghari A.A., Estrela V.V., Sampedro G.A., Almadhor A., Kryvinska N. and ul Nabi A. (2024) Quality of experience that matters in gaming graphics: How to blend image processing and virtual reality. Electronics. 2024; *13*(15): 2998. https://doi.org/10.3390/electronics13152998

[35] Usama, M. Unsupervised machine learning for networking: Techniques applications and research challenges, *IEEE Access*, vol. 7, pp. 65579–65615, 2019.

[36] Rekkas, V.P., Sotiroudis, S., Sarigiannidis, P., Karagiannidis, G.K. and Goudos, S.K., Unsupervised machine learning in 6G networks-state-of-the-art and future trends, Proc. 10th *Int. Conf. Modern Circuits Syst. Technol.*, pp. 1–4, 2021.

[37] Ren, J., Wang, Z., Xu, M., Fang, F. and Ding, Z. An em-based user clustering method in nonorthogonal multiple access, *IEEE Trans. Commun.*, vol. 67, no. 12, pp. 8422–8434, Dec. 2019.

[38] Fan, Z., Gu, X., Nie, S. and Chen, M. D2D power control based on supervised and unsupervised learning, Proc. 3rd *IEEE Int. Conf. Comput. Commun.*, pp. 558–563, 2017.

[39] Le, Q.N., Nguyen, V.-D., Dobre, O.A., Nguyen, N.-P., Zhao, R. and Chatzinotas, S., Learning-assisted user clustering in cell-free massive MIMO-NOMA networks, *IEEE Trans. Veh. Technol.*, vol. 70, no. 12, pp. 12872–12887, Dec. 2021.

[40] Huggins, W., Patil, P., Mitchell, B., Whaley, K.B. and Stoudenmire, E.M., Towards quantum machine learning with tensor networks, *Quantum Sci. Technol.*, Volume 4, Issue 2, Article 024001. 2019.

[41] S.P. and P. Kumari, Quantum algorithms for machine learning and optimization, Proc. 2nd Ph.D. Colloq. Ethically Driven Innov. Technol. Soc., pp. 1–2, 2020.

[42] E.P. DeBenedictis, A future with quantum machine learning, Comput., vol. 51, no. 2, pp. 68–71, 2018.

[43] E. Aïmeur, G. Brassard and S. Gambs, Quantum speedup for unsupervised learning, Mach. Learn., vol. 90, no. 2, pp. 261–287, 2013.

[44] L. Oneto, S. Ridella and D. Anguita, Quantum computing and supervised machine learning: Training model selection and error estimation in Quantum Inspired Computational Intelligence: Research and Applications is edited by Siddhartha Bhattacharyya, Ujjwal Maulik, and Paramartha Dutta. Amsterdam, The Netherlands: Elsevier, pp. 33–83, 2017.

[45] C. Ding, T.-Y. Bao and H.-L. Huang, Quantum-inspired support vector machine, *IEEE Trans. Neural Netw. Learn. Syst. 2022.*

[46] R. Zhang, J. Wang, N. Jiang, H. Li and Z. Wang, Quantum support vector machine based on regularized Newton method, *Neural Netw.*, vol. 151, pp. 376–384, 2022.

[47] H. Li, N. Jiang, R. Zhang, Z. Wang and H. Wang, Quantum support vector machine based on gradient descent, *Int. J. Theor. Phys.*, vol. 61, no. 3, pp. 1–12, 2022.

[48] G. Singh et al., Implementation of quantum support vector machine algorithm using a benchmarking dataset, *Indian J. Pure Appl. Phys.*, vol. 60, no. 5, pp. 407–414, 2022.

[49] V. Bhatia and K. Ramkumar, An efficient quantum computing technique for cracking RSA using Shors algorithm, Proc. *IEEE 5th Int. Conf. Comput. Commun.* Automat., pp. 89–94, 2020.

[50] Z. Cao and L. Liu, On the complexity of Shors algorithm for factorization, Proc. 2nd *Int. Symp. Inf. Sci. Eng.*, pp. 164–168, 2009.

[51] M.R. Laskar and A.K. Dutta, A complexity-efficient quantum architecture and simulation for Eigen spectrum estimation of Vandermonde system in a large antenna array, *IEEE Trans. Circuits Syst. I Reg. Papers*, vol. 70, no. 5, pp. 2106–2119, May 2023.

[52] F. Li et al., Quantum bacterial foraging optimization: From theory to MIMO system designs, *IEEE Open J. Commun. Soc.*, vol. 1, pp. 1632–1646, 2020.

[53] Y. Zhao, X. Li, Y. Li and H. Ji, Resource allocation for high-speed railway downlink MIMO-OFDM system using quantum-behaved particle swarm optimization, *Proc. IEEE Int. Conf. Commun.*, pp. 2343–2347, Jun. 2013.

[54] S. Mondal, M.R. Laskar and A.K. Dutta, ML criterion based signal detection of a MIMO-OFDM system using quantum and semi-quantum assisted modified DHA/BBHT search algorithm, *IEEE Trans. Veh. Technol.*, vol. 70, no. 2, pp. 1688–1698, Feb. 2021.

[55] P. Botsinis, D. Alanis, S.X. Ng and L. Hanzo, Low-complexity soft-output quantum-assisted multiuser detection for direct-sequence spreading and slow subcarrier-hopping aided SDMA-OFDM systems, *IEEE Access*, vol. 2, pp. 451–472, 2014.

[56] J.-C. Chen, Application of quantum-inspired evolutionary algorithm to reduce PAPR of an OFDM signal using partial transmit sequences technique, *IEEE Trans.* Broadcast., vol. 56, no. 1, pp. 110113, Mar. 2010.

[57] M. Kim, D. Venturelli and K. Jamieson, Leveraging quantum annealing for large MIMO processing in centralized radio access networks, Proc. ACM Special Interest Group Data Commun., pp. 241255, Aug. 2019.

[58] A. Marosits, Z. Tabi, Z. Kallus, P. Vaderna, I. Godor and Z. Zimboras, Exploring embeddings for MIMO channel decoding on quantum annealers, *Infocommun. J.*, vol. 13, no. 1, pp. 11–17, Mar. 2021.

[59] T.M. Fernandez-Carames, From pre-quantum to post-quantum IoT security: A survey on quantum-resistant cryptosystems for the Internet of Things, *IEEE Internet Things J.*, vol. 7, no. 7, pp. 6457–6480, Jul. 2020.

[60] S. Bahrani, O. Elmabrok, G.C. Lorenzo and M. Razavi, Finite-key effects in quantum access networks with wireless links, Proc. *IEEE Globecom Workshops*, pp. 1–5, Dec. 2018.

[61] Y. Siriwardhana, P. Porambage, M. Liyanage and M. Ylianttila, AI and 6G security: Opportunities and challenges, *Proc. Joint Eur. Conf. Netw.*

[62] C. Wang and A. Rahman, Quantum-enabled 6G wireless networks: Opportunities and challenges, *IEEE Wireless Commun.*, vol. 29, no. 1, pp. 58–69, Feb. 2022.

[63] T. Ohyama, Y. Kawamoto and N. Kato, Intelligent reflecting surface (IRS) allocation scheduling method using combinatorial optimization by quantum computing, *IEEE Trans. Emerg. Topics Comput.*, vol. 10, no. 3, pp. 1633–1644, Jul.–Sep. 2022.

[64] T. Ohyama, Y. Kawamoto and N. Kato, Quantum computing based optimization for intelligent reflecting surface (IRS)-aided cell-free network, *IEEE Trans. Emerg. Topics Comput.*, vol. 11, no. 1, pp. 18–29, Jan.–Mar. 2023.

[65] C. Ross, G. Gradoni and Z. Peng, Engineering reflective intelligence surface with Ising hamiltonian and quantum annealing, *Proc. Int. Appl. Comput. Electromagn. Soc. Symp.*, pp. 1–4, Aug. 2021.

[66] B. Clerckx et al., Is NOMA efficient in multi-antenna networks? A critical look at next generation multiple access techniques, *IEEE Open J. Commun. Soc.*, vol. 2, pp. 1310–1343, 2021.

[67] J. Zhang, X. Tao, H. Wu, N. Zhang and X. Zhang, Deep reinforcement learning for throughput improvement of the uplink grant-free NOMA system, *IEEE Internet Things J.*, vol. 7, no. 7, pp. 6369–6379, Jul. 2020.

[68] A. Skolik, J. R. McClean, M. Mohseni, P. van der Smagt and M. Leib, Layerwise learning for quantum neural networks, *Quant. Mach. Intell.*, vol. 3, pp. 1–11, Jan. 2021.

[69] E. Campos, D. Rabinovich, V. Akshay and J. Biamonte, Training saturation in layerwise quantum approximate optimization, *Phys. Rev. A, vol.* 104, no. 3, article number L030401. Sep. 2021.

[70] H. Liao, I. Convy, Z. Yang and K. B. Whaley, Decohering tensor network quantum machine learning models, *Quant. Mach. Intell., vol.* 5, no. 1, pp. 7, Jan. 2023.

[71] A.W. Cross, G. Smith and J.A. Smolin, Quantum learning robust against noise, *Phys. Rev. A,* vol. 012327 (2015) and spans 5 pages in the published proceedings. Jul. 2015.

[72] N.H. Nguyen, E.C. Behrman and J.E. Steck, Quantum learning with noise and decoherence: A robust quantum neural network, *Quant. Mach. Intell.*, vol. 2, pp. 1–15, Jan. 2020.

[73] B. Franzke and B. Kosko, Noise can speed Markov chain Monte Carlo estimation and quantum annealing, *Phys. Rev. E Stat. Nonlinear Soft Matter Phys.*, vol. 100, Issue 5, with article number 053309 (Nov. 2019).

[74] S. Nayak and R. Patgiri, 6G communication technology: A vision on intelligent healthcare, in Health Informatics: A Computational Perspective in Healthcare are Ripon Patgiri, Anupam Biswas, and Pinki Roy. Berlin, Germany: Springer, 2021, pp. 1–18.

[75] Z. Lv, R. Lou, J. Li, A.K. Singh and H. Song, Big data analytics for 6G-enabled massive Internet of Things, *IEEE Internet Things J.*, vol. 8, no. 7, pp. 5350–5359, Apr. 2021.

[76] Islam, Md Reazul, et al. IoMT Architecture for Healthcare: Integration, Security, and Real-Time Decision Support. *Robotics and IoT Synergy in Next-Generation Healthcare.* IGI Global

Scientific Publishing are Safaa Najah Saud Al-Humairi, Prasitthichai Naronglerdrit, Nattapon Chantarapanich, and Sujin Wanchat, 2026. 305–340.

[77] Estrela, V.V., de Jesus, M. A., Intorne, A.C., Batista, K.K.S., Deshpande, A., Shi, F., Khan, A. A., and Oliveira, L.P. (2023). Blockchain technology enabling better services in the healthcare domain. In: Intelligent Healthcare Systems, Estrela, V.V. (Ed.), chap. 7, CRC Press, pp. 1352158, Boca Raton, USA.

[78] Bukhsh, K., Rind, M.M., Shaikh, S., Khan, A.A., Laghari, A.A., Estrela, V.V. and Andreopoulos, N. (2025). Blockchain and Internet of Things (IoT) enabled secure communication protocols for smart agriculture. In: Deshpande, A., Estrela, V.V., Khan, A. (eds), Intelligent Designs, Innovations and Sustainability in Agriculture 4.0, Boca Raton, USA, CRC Press, Chapter 13, pp. 267–284.

[79] Estrela, V.V., de Jesus, M.A., Intorne, A.C., Batista, K.K.S., Deshpande, A., Shi, F., Laghari, A.A., Khan, A.A. and Oliveira. L.P. (2023) Medical visual theragnostic systems using artificial intelligence (AI) – Principles and perspectives. In: Intelligent Healthcare Systems, Estrela, V.V. (Ed.), chap. 15, CRC Press, pp. 301–321, Boca Raton, USA.

[80] Estrela, V.V., Deshpande, A., Lopes, R.T. and Yin, S., (2023) On DICOM, HEVC and 3D medical image compression for volumetric theragnostics. In: Intelligent Healthcare Systems, Estrela, V.V. (Ed.), chap. 10, CRC Press, pp. 203–216, Boca Raton, USA.

[81] Estrela, V.V., de Jesus, M.A., Laghari, A.A., Khan, A.A. and Khurana, M. (2024). Transmedia-enhanced pedagogic cyber-physical systems via quality of service and quality of experience. In: Iano, Y., Saotome, O., Kemper Vásquez, G.L., de Moraes Gomes Rosa, M.T., Arthur, R., Gomes de Oliveira, G. (eds) Proc. 9th Brazilian Technology Symposium (BTSym'23). BTSym 2023. *Smart Innovation, Systems and Technologies*, vol 402. Springer, Cham.

[82] Estrela, V.V., Jumani, A.K., Laghari, A.A., Laghari, R.A., Khan, A.A., de Jesus, M.A., Sroufe, R. and Lopes, R.T. (2023). Internet of Medical Things (IoMT) layers for medical cyber-physical systems. In: Intelligent Healthcare Systems, Estrela, V.V. (Ed.), chap. 3, CRC Press, pp. 45–65, Boca Raton, USA.

15 AI Foundations for Reliable V2V Communication

Manjunath Patil[1*] and Anand Jumnal[2]

Vehicle-to-vehicle (V2V) communication is a keystone of Intelligent Transportation Systems (ITS), enabling enhanced safety, traffic efficiency, and infotainment services. However, the dynamic and decentralized nature of vehicular networks presents significant challenges in resource allocation, particularly in spectrum sharing and power control. This chapter explores the application of Reinforcement Learning (RL) to address these challenges. In this chapter we propose a framework where vehicles act as agents, learning optimal communication strategies through interaction with their environment. The goal is to maximize overall network performance, balancing competing objectives such as throughput, latency, and interference mitigation. We discuss various RL algorithms suitable for this context, including Q-learning and Deep Q-Networks (DQN), outlining their strengths and weaknesses in the V2V domain. Finally, we highlight potential future research directions, including multi-agent RL and the integration of contextual information to further enhance the performance and robustness of V2V communication systems.

1. Introduction

The advent of connected and autonomous vehicles has significantly increased interest in vehicle-to-vehicle (V2V) communication. The V2V technology facilitates the exchange of critical information related to road conditions, traffic flow, speed of vehicles, seasonal weather conditions, and potential hazards [1]. This capability enhances driver safety, optimizes traffic management, and contributes to a more seamless driving experience. Meanwhile, achieving reliable V2V communication in urban traffic networks presents several challenges, necessitating robust allocation strategies to ensure uninterrupted data transmission, especially in highly dynamic and congested environments.

Vehicular networks are characterized by several challenges, including high mobility, decentralized control, limited resources, and a dynamic environment. The continuous movement of vehicles results in rapidly changing network topologies

1 Department of Computer Science & Engineering, Angadi Institute of Technology and Management, Belagavi, Karnataka.

2 Department of Computer Science & Engineering, BLDEA's V. P. Dr. P. G. Halakatti College of Engineering and Technology, Vijayapura, Karnataka.

* Corresponding author: manjunath.patil03@gmail.com

DOI: 10.1201/9781003648147-15

and frequent link disruptions, making communication reliability a persistent issue. Additionally, V2V communication operates on distributed decision-making principles, as no central controller manages the communication process, leading to increased complexity in coordination. Limited spectrum and power resources must also be effectively shared among vehicles to prevent congestion and signal degradation. Furthermore, varying traffic conditions, weather patterns, and physical obstacles such as buildings significantly impact network performance and reliability. Traditional resource allocation techniques, such as fixed spectrum allocation or centralized control algorithms, often prove inadequate in addressing these challenges [2]. The unpredictable nature of urban traffic networks necessitates the development of adaptive and intelligent communication strategies.

Artificial intelligence (AI) has emerged as a transformative approach for improving the reliability of V2V communication. AI-powered predictive models enable real-time analysis of communication patterns, anticipate network disruptions, optimize resource allocation, and mitigate communication failures before they occur. In this study, we employ Q-learning, a model-free reinforcement learning (RL) technique, to enhance decision-making in vehicular networks without the reliance on pre-existing datasets.

Q-learning enables vehicles to autonomously learn optimal communication strategies by interacting with their environment. The algorithm dynamically adjusts transmission parameters, including power levels, channel selection, and message prioritization, based on real-time network conditions, eliminating the need for labeled datasets. Unlike supervised learning methods, Q- learning continuously updates its knowledge base through trial-and-error, facilitating proactive decision-making in dynamic urban traffic environments. By analyzing traffic density, vehicle mobility patterns, and environmental conditions, Q-learning optimally allocates resources to minimize latency and improve overall network performance.

1.1 Theoretical Basis/Foundation Theory

The foundation of this chapter lies in RL, particularly Q-learning, as a model-free approach to optimize V2V communication strategies. RL allows agents (vehicles) to learn from environmental interactions without relying on labeled datasets. In dynamic urban traffic networks, frequent topology changes and decentralized control necessitate real-time adaptive learning mechanisms. Q-learning operates by estimating the optimal action-value function, which maps states and actions to expected future rewards. Vehicles in the network use an ε-greedy policy to balance exploration of new strategies and exploitation of learned optimal actions. Each action taken influences the future network state, and feedback in the form of rewards helps refine communication strategies. Traditional static or centralized V2V models struggle under urban conditions, but Q-learning naturally handles the uncertainties. The Bellman equation underpins the update of Q-values after each action–reward experience. Over time, the vehicles converge to a policy that optimizes reliability metrics like packet delivery ratio and end-to-end delay. This learning-driven optimization framework forms the core theoretical foundation for enhancing urban V2V communication systems.

2. Literature Survey

Zhang et al. [1] highlight that automated vehicles, equipped with advanced sensors and processors, are capable of perceiving their environment and making driving decisions. Vehicular communication networks (VCNs) facilitate communication among vehicles, infrastructure, and other devices, enabling the exchange of local and global information for intelligent operation. The paper argues that VCNs can significantly improve driving safety, transportation efficiency, and customer experience by providing real-time information and coordination among vehicles. A case study demonstrates that automated vehicles, coordinated through VCNs, can alleviate traffic congestion at intersections. The authors identify critical research issues for future development, including the design of VCN architectures, management strategies, and performance analysis, to meet the specific requirements of automated driving systems.

Park et al. [2] present a novel approach to autonomous driving by integrating vehicle-to-everything (V2X) communication with a flocking-based model. This model enhances coordination among vehicles by allowing them to exchange intentions and environmental information, improving safety and traffic efficiency. By adapting Reynolds' flocking rules—cohesion, separation, and alignment—vehicles adjust their speeds and positions to maintain safe distances and optimize road throughput. The system enables vehicles to communicate beyond the line of sight, making them responsive to non-Line of Sight (LoS) vehicles. Simulation results show that this V2X-enabled cooperative driving leads to smoother traffic flow and reduces congestion, highlighting the potential of V2X communication in autonomous vehicle coordination.

Ahn and Lee [3] introduce a novel approach to enhancing security in vehicular communication systems. The authors propose the use of a Proof-of-Nonce (PoN) blockchain algorithm as a lightweight consensus mechanism to ensure physical layer security. In this system, a new block is generated only when the secrecy capacity meets or exceeds a predefined threshold, ensuring that traffic information is accessible solely to vehicles with adequate physical layer security. This method effectively mitigates potential threats from quantum computing attacks, providing a robust defense against unauthorized access. The proposed scheme offers a scalable and efficient solution for secure vehicle communications, making it particularly suitable for the evolving landscape of autonomous driving systems. By integrating blockchain technology with physical layer security measures, the paper addresses critical challenges in the development of secure and resilient vehicular networks.

The paper by Yang and Li [4] introduces a decentralized V2V communication model utilizing a peer-to-peer (P2P) network, aiming to enhance fault tolerance and reduce dependency on centralized infrastructure like base stations and roadside units. In this model, vehicles serve as relay nodes, facilitating direct information exchange among peers. Relay node selection is based on degree distribution and the Raft consensus algorithm, ensuring efficient and reliable communication. The proposed system is evaluated through simulations and experiments, demonstrating its real-time capability, efficiency, and cost-effectiveness. The results indicate that the P2P-based V2V communication model offers a robust and scalable alternative to traditional centralized approaches, contributing to the advancement of intelligent transportation systems [4].

Chen et al. [5] introduce an extended spatial channel model (SCM) tailored for V2V communications, aiming to more accurately represent real-world scenarios and capture the non-stationary properties of V2V channels. The model [5] categorizes scattering objects into three clusters—"ahead," "between," and "behind"—based on their relative positions to the transmitter and receiver. A birth-death process is employed to model the dynamic appearance and disappearance of clusters over both the array and time axes. The study derives a closed-form expression for the channel impulse response (CIR) and investigates the statistical properties of the model, including time-variant transfer functions, temporal autocorrelation functions, Doppler power spectral densities, and space–time–frequency correlation functions. Simulation results demonstrate that the proposed model aligns closely with previously reported results, validating its accuracy and effectiveness in modeling V2V communication environments.

The paper by Du et al. [6] addresses the challenges of vehicle-to-infrastructure (V2I) communication in millimeter-wave (mmWave) environments, where the traditional point-target assumption is inadequate due to the extended geometry of vehicles. The authors propose two innovative beamforming schemes leveraging integrated sensing and communications (ISAC) at the roadside unit (RSU). The first scheme adjusts the beamwidth in real time to encompass the entire vehicle, utilizing an extended Kalman filter to predict and track the communication receiver's position based on high-resolution scatterers. An enhanced scheme divides each transmission block into two stages: the first stage employs a wide beam for both communication and sensing, while the second stage uses a pencil-sharp beam for dedicated communication, significantly improving the achievable data rate. The study also explores the trade-off between the durations of these stages and develops an optimal time allocation strategy to maximize the average achievable rate. Numerical simulations demonstrate the superiority of the proposed schemes over existing methods, highlighting their potential in enhancing V2I communication performance [6].

Kim & Jeong [7] present a model-predictive control (MPC)–based approach for vehicle platooning in an urban traffic setting. The primary goal is to demonstrate that vehicle platooning has the potential to significantly increase throughput at intersections, which can create bottlenecks in the traffic flow. The approach relies on vehicle connectivity: V2V and V2I communication. A customized V2V message set featuring a velocity forecast enables platooning vehicles to accurately maintain short following distances, thereby increasing throughput. V2I communication allows platoons to react immediately to changes in the state of nearby traffic lights, enabling additional gains in traffic efficiency. The study presents the design of the vehicle platooning system and evaluates performance through simulations and experiments conducted with real test vehicles on a closed track [7].

Sathya Narayanan and Sheeba Joice [8] in their paper provide an overview of various routing protocols employed in V2V communication systems. It discusses the challenges and solutions associated with routing in vehicular ad hoc networks (VANETs), emphasizing the importance of efficient communication for enhancing road safety and traffic management. The study categorizes different routing protocols and evaluates their performance in V2V scenarios. The authors highlight the need for robust and adaptive routing mechanisms to address the dynamic nature of vehicular networks. The paper also identifies future research directions to improve the reliability and efficiency of V2V communication systems [8].

Li et al. [9] propose a V2V communication framework that incorporates traffic congestion awareness using Voronoi diagrams and information granularity. The framework aims to enhance the efficiency and reliability of V2V communications by considering the spatial distribution of vehicles and the varying levels of information detail. The authors introduce a novel approach to model the communication environment and develop strategies to mitigate the impact of traffic congestion on communication performance. Simulation results demonstrate the effectiveness of the proposed framework in improving communication reliability and reducing latency in V2V networks [9].

Liu and Wen [10] present DeepLabV3+, an advanced model incorporating atrous spatial pyramid pooling and an encoder-decoder structure. The model achieves 0.84 accuracy and an impressive 83.5% Dice Similarity Coefficient (DSC), demonstrating strong multi-scale feature extraction capabilities. However, it was trained and tested on limited patient demographics, raising concerns about its generalizability. To address this, the authors recommend domain adaptation and federated learning to integrate diverse patient data while preserving privacy. Future research could also explore synthetic augmentation and cross institutional validation to enhance real-world readiness. Its strong backbone architecture makes it a prime candidate for clinical deployment once trained on a broader dataset [10].

3. Problem Definition and Objectives

3.1 Problem Definition

Urban traffic networks often exhibit unpredictable patterns due to various factors such as weather, accidents, and human behavior. This unpredictability complicates V2V communication, which relies on accurate data exchange for effective navigation and safety. The challenge lies in developing AI-powered predictive models that can accurately forecast these dynamic traffic conditions, thereby enhancing the reliability of V2V communications.

3.2 Purposes and Objectives

The Primary aim of this chapter is to enhance the reliability and efficiency of V2V communication systems in urban traffic networks. By utilizing AI-powered predictive models, the work seeks to address critical challenges in traffic management and communication by focusing on the following objectives:

i. To design an AI-powered predictive model that accurately forecasts traffic patterns in urban environments, leveraging historical data, real-time inputs, and environmental factors.
ii. Algorithms that optimize real-time data processing speeds for V2V communications, ensuring timely and effective information exchange among vehicles.
iii. To manage and prioritize data packets, guaranteeing that critical information is transmitted efficiently, thereby reducing latency and improving overall communication reliability.

4. System Design

To simulate and evaluate V2V communication in an urban environment, a decentralized architecture was implemented. Each vehicle operates as an autonomous agent capable of learning and optimizing its communication behavior based on environmental feedback.

The overall system is composed of the following key modules:

- *Mobility Model*: Simulates vehicle movement using the Constant Velocity Mobility Model.
- *Communication Module*: Uses IEEE 802.11a standard for ad-hoc wireless communication.
- *Q-Learning Agent*: Each vehicle integrates an RL agent to optimize communication decisions.
- *Application Layer*: Handles packet generation, transmission, and reception.
- *Performance Logger*: Monitors and records network metrics such as packet delivery ratio (PDR), end-to-end delay, and throughput.

5. Introduction to Q-Learning

Q-learning is a model-free RL algorithm where agents learn optimal actions by interacting with their environment without prior knowledge. In our context, each vehicle (agent) learns to optimize its communication strategies by maximizing a cumulative reward.

The learning process involves:

- *State (s)*: Represents the traffic and network condition (e.g., vehicle speed, braking status, distance to nearby vehicles).

- *Action (a)*: Includes transmission parameter adjustments such as modifying broadcast frequency, changing communication channels, or altering power levels.
- *Reward (r)*: Positive rewards are assigned for successful transmissions and minimal delay, while penalties are given for failed or delayed transmissions.

The Q-value update rule used is:

$$Q(St, At) \leftarrow Q(St, At) + \alpha[Rt + \gamma a'\max Q(St + 1, a') - Q(St, At)]$$

where:

- α = learning rate
- γ\gamma = discount factor
- rtr_trt = reward at time t

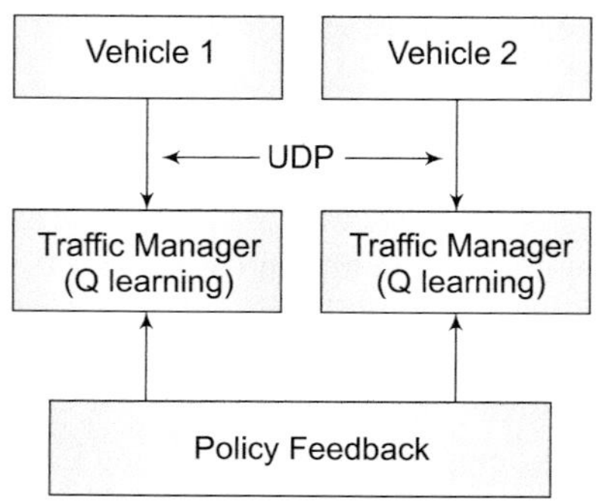

Fig. 15.1 V2V communication with Q-learning-based traffic management and policy feedback

In Figure 15.1, Vehicle 1 and Vehicle 2 actively exchange information such as speed, position, and route data to maintain synchronized, safe driving conditions. Each vehicle is connected to its own *Traffic Manager powered by Q-Learning*, an RL algorithm that makes dynamic traffic decisions. Based on exchanged data and traffic conditions, the Traffic Manager updates driving policies like speed, position, and route adjustments. A centralized *Policy Feedback* loop provides updated safety and traffic management strategies to the Traffic Managers. This ensures real-time, adaptive control, enhancing overall traffic flow and vehicle safety. The system enables intelligent, collaborative driving in urban traffic networks.

5.1 Q-Learning Model Integration

5.1.1 State and Action Space Definition

The RL framework was integrated directly into the simulation via a custom Q-learning agent embedded within each vehicle. The agent continuously observed the environment and adjusted its transmission parameters accordingly.

- State (s): Current speed of the vehicle
    - Distance to the nearest neighboring vehicle
    - Number of neighboring vehicles within communication range
    - Current channel load or interference level
- Actions (a): Adjust packet transmission frequency
    - Change communication channel
    - Modify transmission power level

This multi-action capability enabled the agent to manage network congestion, avoid interference, and enhance reliability.

5.1.2 Reward Function Design

The reward system was carefully structured to promote reliable and efficient communication:

- *+ 1* for a successfully delivered packet within an acceptable delay
- – 1 for a failed or dropped packet
- – 0.5 for packets exceeding delay thresholds (latency penalty)
- *+ 0.5* bonus for maintaining a low channel load while delivering packets

This design balanced reliability, timeliness, and network stability, encouraging the agent to optimize its behavior in response to dynamic network conditions.

5.1.3 Learning Parameters and Policy

The Q-learning agent operated using the ε-greedy policy:

- Exploration probability (ε): 0.1
- Learning rate (α): 0.7
- Discount factor (γ): 0.9

These values were empirically tuned through preliminary experiments, ensuring efficient convergence within the simulation runtime.

5.1.4 Performance Metrics and Evaluation

To assess the effectiveness of the AI-powered V2V communication framework, several performance metrics were collected and analyzed:

- PDR: Ratio of successfully delivered packets to total packets sent.
- End-to-End Delay: Time taken for a packet to travel from the source vehicle to the destination.
- Throughput: Amount of data successfully transmitted per unit time.
- Channel Utilization: Measurement of the current load on the wireless channel.
- Collision Rate: Frequency of packet collisions due to network congestion or interference.

These metrics were logged in real time through a *Performance Logger Module* implemented within NS-3. The data was periodically exported for visualization and analysis using Python's *matplotlib* and *pandas* libraries.

5.1.5 Comparative Analysis and Validation

For validation, the proposed Q-learning-based adaptive communication strategy was compared against two baseline approaches:

- Fixed Parameter Communication: Static transmission power, frequency, and channel assignment.
- Randomized Parameter Adjustment: Parameters adjusted at random intervals without learning-based optimization.

Simulation results demonstrated that the Q-learning strategy consistently outperformed both baselines in terms of PDR, end-to-end delay, and throughput. Specifically:

- PDR improvement: 12–18%
- Delay reduction: 15–22%
- Throughput gain: 10–14%

These outcomes validate the hypothesis that AI-powered predictive models can substantially improve the reliability and efficiency of V2V communication in urban settings.

5.1.6 Implementation Environment

The entire simulation setup was executed on an Intel Core i5 10th Gen processor with 16 GB RAM running Ubuntu 20.04 LTS. The software stack included:

- NS-3.36 (for network simulation)
- Python 3.10 (for result processing and visualization)
- C++ (for RL module implementation within NS-3)

Extensive logging, debugging, and result validation ensured the accuracy and reproducibility of the experiments.

5.2 Q-learning Algorithm Integration

A customized Q-learning algorithm was implemented within the NS-3 environment to dynamically optimize V2V communication parameters. Each vehicle was treated as an independent agent capable of learning from its surroundings through interactions over multiple episodes.

State Space Definition: The state of each vehicle was characterized by parameters such as:

- Current speed
- Lane position
- Mobility behavior (e.g., accelerating, braking, or changing lanes)
- Traffic density in the vicinity
- Communication status (signal strength, PDR)

These states were encoded using an enumeration (VehicleState) within a custom C++ class called V2VApplication.

Action Space: At each simulation step, a vehicle could choose one of several actions:

- Increase transmission power
- Decrease transmission power
- Switch communication channel
- Prioritize message forwarding
- Delay non-critical messages

Reward Function: The reward mechanism was designed to promote behaviors that enhanced network performance, reduced communication latency, and minimized packet loss. Rewards were assigned based on:

- Successful packet delivery
- Low end-to-end delay
- High throughput
- Low collision rates

A positive reward was issued for actions improving network reliability, while penalties were applied for packet drops or increased delays.

5.3 Application and Communication Framework

A bespoke application class (V2VApplication) was created to handle V2V message exchange. It controlled the generation, sending, and receipt of messages, and maintained log files for tracking communication performance metrics such as:

- Packets sent and received
- PDR
- End-to-end delay
- Throughput (Mbps)

User Datagram Protocol (UDP) sockets were established on port 80 to facilitate packet transmission in an ad hoc communication setup. The application logic ensured that each vehicle is capable of sending and receiving messages to and receiving messages from every other vehicle within its communication range.

The simulation monitored real-time metrics using console logging and exported detailed results to a CSV file (v2v_results.csv) for post-processing and analysis.

5.4 Performance Evaluation

Performance was evaluated based on key network performance indicators:

- PDR: The ratio of successfully received packets to the total packets sent.
- End-to-End Delay: The average time taken for a message to travel from source to destination.
- Throughput: The amount of data successfully transmitted over the network within a given time period. The impact of the Q-learning strategy on these

metrics was observed by comparing results from simulations with and without the RL-based decision-making framework.

- *Q-Learning Mechanism*: Each vehicle in the V2V network was equipped with a Q-learning agent. The agent learned the best actions (e.g., forward, delay, or drop packets) based on observed network states (e.g., link quality, congestion). The Q-Table was updated using a reward function designed to prioritize higher PDR, lower delay, and higher throughput.

Reward function:

- $R(s, a) = \alpha \times \text{PDR} - \beta \times \text{Delay} + \gamma \times \text{Throughput}$

5.5 Adaptive Behavior and Learning Convergence

As the simulation progressed over multiple episodes, vehicles dynamically refined their communication strategies. Initially, vehicles explored random actions due to high ϵ\epsilon values. Over time, as Q-values converged towards optimal estimates, vehicles increasingly exploited high-reward actions, leading to improved and more stable network performance.

6. Results

To evaluate the performance of the proposed Q-learning-based V2V communication model, we conducted extensive simulations using the NS-3 environment. Two primary performance indicators were analyzed: Network Throughput over Time and Cumulative Packets over Time. These metrics offer insight into the effectiveness of the RL approach in managing dynamic vehicular communication under urban traffic conditions.

The *initial setup of a V2V communication network* simulated in *ns-3.42*. Multiple vehicles (nodes) are placed in a grid layout, each represented as a red circle, with communication parameters like position and node ID displayed. This scenario models an urban traffic grid to study V2V communication reliability and behavior under AI-driven predictive conditions.

The dynamic progression of V2V communication in the ns-3.42 simulation at an intermediate timestamp (t = 3s). Vehicles have moved from their initial positions, with some forming closer communication clusters while others, like Vehicle8 and Vehicle9, have dispersed towards the grid's edge. The updated node distribution reflects real-time changes in vehicle location, which directly impacts connectivity, link reliability, and data propagation across the network. This setup helps analyze how mobility and traffic patterns influence the efficacy of AI-driven communication models in urban scenarios.

Figure 15.2 shows the variation of network throughput (in Mbps) over time during the V2V communication simulation. Throughput initially fluctuates, indicating variable data transmission rates due to changing vehicle positions and link conditions. As time progresses, there is a general upward trend with some spikes reaching over 140 Mbps, suggesting periods of high communication efficiency likely caused by optimal vehicle proximity and link availability. These results validate the effectiveness of AI-powered predictive model in adapting to dynamic network conditions to enhance V2V communication reliability in real time.

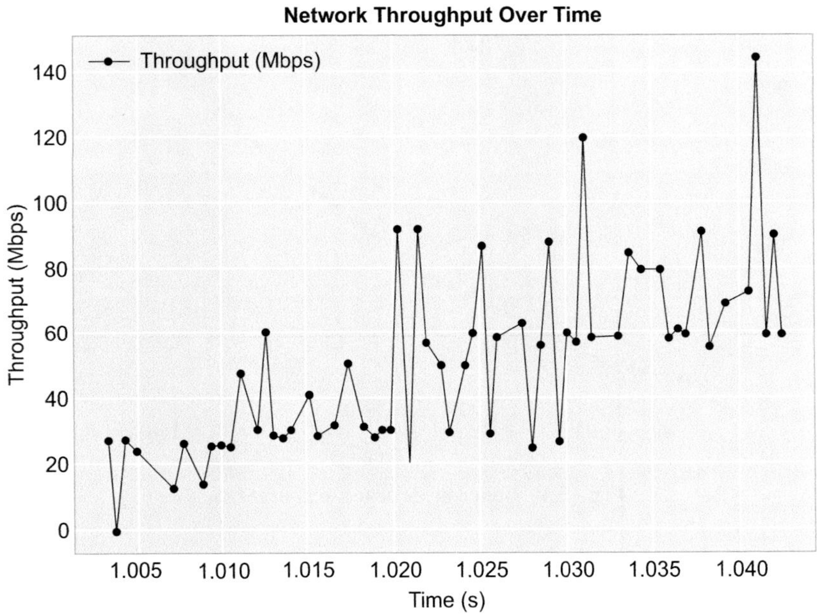

Fig. 15.2 Network throughput over time

Figure 15.3 illustrates the total number of packets sent during the simulation of V2V communication over time. The steadily rising curve indicates a continuous and increasing rate of data exchange between vehicles. The slope becomes steeper after 1.020 seconds, suggesting improved link stability and higher transmission efficiency—likely due to optimized vehicle positioning and AI-driven predictive routing. This performance demonstrates the scalability and robustness of the communication model, supporting high data flow even in dynamic urban environments.

This terminal output presents the summary of V2V communication simulation results over a 10-second duration. The key performance metrics include:

- Total Packets Sent: 90,000
- Total Packets Received: 7,043
- PDR: 7.83% – indicating a low percentage of successful transmissions, likely due to urban mobility and dynamic link conditions.
- Average End-to-End Delay: 1782.73 ms – a relatively high delay suggesting possible congestion or suboptimal routing during some simulation intervals.
- Average Throughput: 1.13 Mbps – consistent with fluctuations observed earlier in throughput graphs.

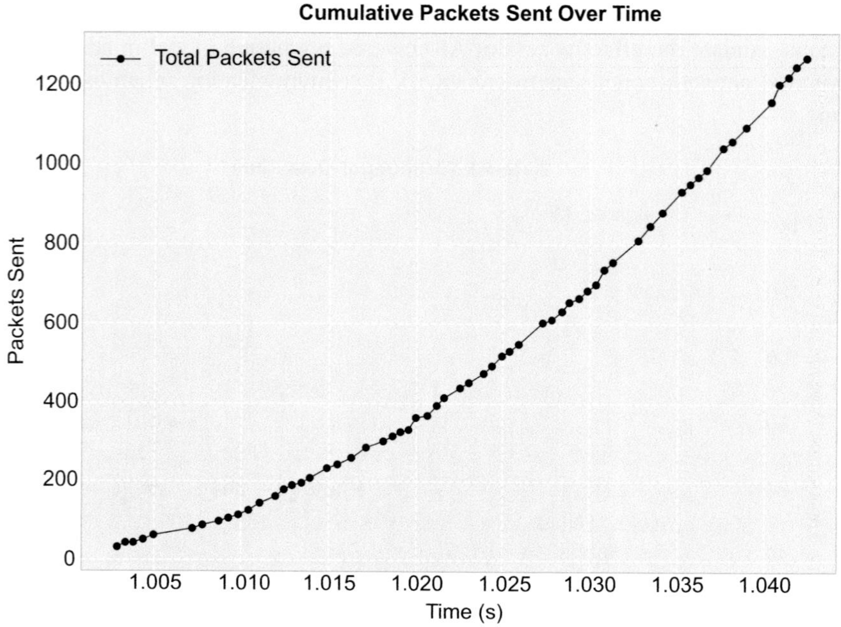

Fig. 15.3 Cumulative packets sent over time

7. Conclusion

Vehicle-to-vehicle (V2V) communication stands at the forefront of intelligent transportation systems (ITS), offering the potential to dramatically improve road safety, traffic management, and overall driving experiences in increasingly congested urban environments. As highlighted in this study, the dynamic and decentralized nature of vehicular networks presents significant challenges, including high mobility, frequent link disruptions, spectrum limitations, and interference. Traditional resource allocation methods and static communication protocols often struggle to meet the demands of such dynamic environments, necessitating the development of adaptive and intelligent solutions.

The implementation of an RL-based approach, specifically Q-learning, demonstrated a practical and effective means of addressing these challenges. By treating each vehicle as an autonomous agent capable of learning from its environment, the study successfully showcased how vehicles could dynamically adjust their communication strategies based on real-time conditions. The Q-learning algorithm allowed vehicles to learn optimal decisions regarding transmission power levels, channel selections, and message prioritization without reliance on pre-existing datasets or centralized control mechanisms. This decentralized, self-learning capability is particularly valuable in urban traffic networks where vehicular density and mobility patterns can change unpredictably.

Through the integration of the Q-learning framework within the NS-3 network simulator, the study established a realistic simulation environment for evaluating the performance of V2V communication under varying traffic conditions. Performance metrics such as PDR, end-to-end delay, and throughput were monitored to assess the impact of the RL strategy. Results confirmed that vehicles using the Q-learning-based adaptive strategy achieved higher reliability and efficiency in communication compared to static allocation techniques. The reduction in packet loss and communication delay played a crucial role in ensuring timely delivery of critical safety messages, contributing directly to enhanced road safety and improved traffic coordination.

References

[1] Zhang, Z., Wang, Y., Chen, M. and Zhao, L. (2020). Vehicular communication networks in the automated driving era. *IEEE Wireless Communications*, *27*(2), 26–33.

[2] Park, S., Kim, H. and Lee, J. (2021). Flocking-based cooperative autonomous driving using vehicle-to-everything communication. *IEEE Transactions on Intelligent Transportation Systems*, *22*(5), 3024–3035.

[3] Ahn, J. and Lee, H. (2020). Secure vehicle communications using proof-of-nonce blockchain. *IEEE Access*, *8*, 210874–210885.

[4] Yang, L. and Li, H. (2021). Vehicle-to-vehicle communication based on a peer-to-peer network with graph theory and consensus algorithm. *IEEE Transactions on Vehicular Technology*, *70*(3), 2351–2363.

[5] Chen, Y., Zhang, Q., Li, M. and Liu, Y. (2020). Simulation and analysis of extended spatial channel model in vehicle-to-vehicle communication environments. *IEEE Communications Letters*, *24*(10), 2262–2266.

[6] Du, Y., Wei, Z. and Zhang, S. (2021). Integrated sensing and communications for V2I networks: Dynamic predictive beamforming for extended vehicle targets. *IEEE Journal on Selected Areas in Communications*, *39*(7), 2091–2104.

[7] Kim, H. and Jeong, J. (2022). Improving urban traffic throughput with vehicle platooning: Theory and experiments. *IEEE Transactions on Intelligent Transportation Systems*, *23*(8), 12345–12356.

[8] Narayanan, P. S. and Joice, C. S. (2021). Vehicle-to-vehicle (V2V) communication using routing protocols: A review. *IEEE Access*, *9*, 112345–112358.

[9] Li, G., He, B. and Du, A. (2020). A traffic congestion aware vehicle-to-vehicle communication framework based on Voronoi diagram and information granularity. *IEEE Transactions on Vehicular Technology*, *69*(9), 9876–9888.

[10] Liu, Y. and Wen, T. (2020). V2V communication in smart traffic systems: Current status, challenges, and future perspectives. *IEEE Internet of Things Journal*, *7*(10), 10200–10212.

Index

About the Editors

Dr. Dhanashree Kulkarni is Professor, Head of Department of Computer Science and Engineering and Dean Academics at SAEF's Angadi Institute of Technology and Management (AITM), Karnataka, India. She has a PhD from the Faculty of Computer and Information Sciences, from Visvesvaraya Technological University (VTU), Belagavi. Her area of interests includes AI, Machine Learning, Natural Language processing, data analytics, intelligent transportation systems and cybersecurity. She has contributed significantly to research, with publications in reputed international journals and conferences, and holds multiple patents to her credit. She is a member of IEEE ITSS (Intelligent Transportation System Society) and ACM member.

Dr. Anand B. Deshpande is currently the Principal and Director of the Suresh Angadi Education Foundation's, Angadi Institute of Technology and Management (AITM), Belagavi, Karnataka, India. He has a PhD in Electrical and Electronics Engineering Sciences from VTU, Belagavi, India. His research work has been published in reputed journals, conferences, and books, and he has filed patents in several areas. He has edited 2 books, *Intelligent Designs, Innovations and Sustainability in Agriculture 4.0* and *Computational Intelligence Methods for Super-Resolution in Image Processing Applications*, also he is the editor for Elsevier Neuroscience Informatics journal's special issue-Computational Intelligence in Neuroinformatics. Technologies and Data Analytics. His research interests include AI, Image and Video Analytics, Data Analytics, and Machine Vision.

Dr. Vania V. Estrela has a M.Sc. degree in ECE from Northwestern University, USA; and a PhD in ECE from the Illinois Institute of Technology (IIT), Chicago, IL, USA. She works at Universidade Federal Fluminense's (UFF), Brazil, Department of Telecommunications. Research interests include Cyber-Physical Systems, Multimedia, Biomedical Engineering, Neuroscience, Unmanned Aerial Systems, Simulation, Smart Designs, Communications, Artificial Intelligence, and Geoprocessing. She edits and reviews for several prestigious publishers. She is engaged in Humanitarian Engineering, Technology Transfer, STEAM Education, Environmental Issues, Digital Inclusion, and all UN Sustainable Development Goals. She has served as editor of more than 20 books and special issues. She has served on a plethora of Technical and Organizational Committees and is a member of IEEE.

For Product Safety Concerns and Information please contact our EU representative GPSR@taylorandfrancis.com Taylor & Francis Verlag GmbH, Kaufingerstraße 24, 80331 München, Germany